養好肺‧
強體質

64 道增強肺臟日常食療

編者話

　　病菌無處不在，吸入氣管及肺部，容易引起咳嗽不適等呼吸道及氣管毛病，削弱肺臟健康。尤其秋季乾燥之時，皮膚、口鼻、咽喉及肺臟更易受秋燥影響，肌膚繃緊、唇乾舌燥，身體臟腑同樣受氣候變化而有變。

　　根據中醫理論，外邪之物如何傷害肺臟？怎樣透過四季養生達到潤肺止咳之效？本書特別邀請資深註冊中醫師譚莉英撰文詳釋，以專業的中醫角度，為讀者深入解釋增強肺部之根本要素，防範病毒侵襲。

　　書內提供 64 道日常食療，將潤肺養肺的食材應用於湯水、家常小菜、粥品、茶飲或甜品，除了有止咳潤肺、增強抗病力的湯水，只要懂得配搭合適的強肺材料，靈活變化，小炒、甜點也同具調理身體、增強體質之目的。

　　防範病毒感染、消除病毒，就讓我們從此刻開始！

目錄

滋潤湯水

家常菜 • 粥

潤肺甜點

清潤茶飲

中醫論潤肺養生法

註冊中醫師譚莉英

千草醫藥坊創辦人及主診醫
師、香港註冊中醫師、香港
大學專業進修學院客席講
師、專欄作家

　　《本草綱目》記載：「治療腹脹心痛，利大小便，補中益氣，溫肺止咳，安心定膽益志，養五臟，治產後血暈。」

　　《陰陽平衡養生祛病》指出：「肺主氣，為藏魄之處。」

　　上天公平的賜予我們五臟六腑，心、肝、脾、肺、腎稱之為五臟。而五臟具有化生和貯藏精氣的共同生理功能，同時又各有專司，而且軀體官竅有着特殊的聯繫，形成了以五臟為中心的系統，而心得生理功能起主宰的作用。肺在五臟中所居位置最高，喜潤惡燥，容易被外邪侵襲，故稱為嬌臟。六淫邪氣的風、寒、暑、濕、燥、火之中，肺最怕燥邪。燥邪侵肺，可導致肺的陰津不足，除可見到鼻乾咽燥之外，還可見到乾咳少痰等徵狀。中醫五臟的養生觀念來看，肺在五行之中屬陰，所以秋天的燥熱氣最傷肺，此時以補肺和清燥熱為最好時機，從而達致養生功效。

中醫對肺功能的理解

肺主氣——肺是主管呼吸之氣和一身之氣。

首先，肺主呼吸之氣，是指通過呼吸運動，吸入自然界的清氣，呼出濁氣，實現了體內外氣體交換。中醫學認為，呼吸其實不單靠肺氣來完成，還需有賴於腎的協作。肺為氣之主，腎為氣之帥。肺主呼，腎主納，一呼一納，即一出一入，才完成呼吸運動。當肺司呼吸功能正常，則氣道通暢，呼吸調勻。若病邪犯肺，則影響其呼吸功能，導致出現喘促、呼吸不利等症。

其次肺主一身之氣，即肺能主持、調節全身各臟腑之氣的作用，通過呼吸而參與氣的生成和調節氣機的作用。

肺主水——肺是宣發和肅降對身體內水液的輸送、運行和排泄的疏通和調節作用。

人體內的水液代謝，是由肺、脾、腎、小腸和大腸及膀胱等臟腑共同完成。肺氣宣發分為兩類：一使水液迅速向上向外輸至全身，外達皮毛。二使經肺代謝後的水液被身體利用後的廢水和剩餘的水分，通過呼吸、皮膚的汗孔蒸發而排出。這些亦是肺在調節水液代謝中的作用，也就是肺的通調水道的生理功能。當肺氣宣降失調，會失去行水的能力，造成水道不通，則會出現水液輸布和排泄障礙，引起痰飲、水腫等問題。

肺主治節——指肺輔助心臟治理調節全身氣、血、津液及臟腑生理功能的作用。

心為君主之官，為五臟六腑之大主。肺的治節作用，主要體現於四方面，包括肺主呼吸、調節氣機、助心行血、宣發和肅降。

肺主宣肅——肺通過呼吸運動吸入自然界的清氣，宣發出濁氣，宜宣宜肅來完成清呼濁、吐納新的作用。

肺也可作為通調水道，肺為水之上源，肺氣肅降則能通調水道，使水液代謝產物下輸膀胱。

中醫常稱肺為嬌臟，意思是指肺臟清虛嬌嫩而容易受外邪入侵。肺氣與秋氣相應，肺氣旺於秋，肺與秋季、西方、燥、金、白色、辛味等有內在的的聯繫。燥氣當令，此時的燥邪極易侵犯人體而耗損肺之陰津，導致出現乾咳、皮膚和口鼻乾燥等症狀。

四季潤肺健肺之法

春天

由於春霧潮濕，寒氣未過，適逢轉風天，風、寒、濕邪易襲人體，所以補肺護肺首選是——補氣防寒、驅風、溫陽化濕。

藥食方面：可多進食北芪、藿香、白朮等。

保健方面：可以針灸。針刺後在穴位上佐以艾灸，有助溫經通絡、行氣活血、驅風化濕，此法老幼皆宜。

食療方面：

百合粥（二人份量）

材料：乾百合 3/4 兩，白米 1.5 兩，海鹽適量。

做法：將百合與米分別淘洗乾淨，放入鍋中加水，用小火煨煮。待百合與米熟爛時，加適量鹽調味，即可食用。

功效：對老年人及久病後身體虛弱而有心煩失眠、易怒者尤為適宜。如在百合粥加入甜杏仁 2 錢同煮，即成百合杏仁粥，適宜於肺陰虧虛之久咳、乾咳無痰、氣逆微喘等患者食用。

宜忌：孕婦和長期病患者，服用前先諮詢合資格的註冊中醫師。

夏天

暑濕之邪，容易令肺熱生痰濕，所以夏天補肺護肺著重清肺熱和化痰祛濕。

食療方面：

清熱綠豆湯（二人份量）

材料：綠豆3/4兩，蓮子半兩，
　　　臭草半兩，冰糖適量。

做法：將綠豆用清水浸泡2小時以上；
　　　蓮子與臭草洗淨備用。將泡好的綠豆加水用
　　　大火煮沸，至綠豆開花後轉小火，加入蓮子
　　　繼續熬煮，煮至豆開嫩滑時，加入冰糖和臭
　　　草續煮，待冰糖完全溶化即可關火食用。

功效：綠豆清熱解毒；蓮子養心去火；臭草清熱解
　　　毒，搭配食用可以去除肺火。

宜忌：小孩份量減半。孕婦、產後、老人、大病、
　　　久病和長期病患者，服用前先諮詢合資格的
　　　註冊中醫師。尿糖高出正常水平者要注意冰
　　　糖的份量。

秋天

秋冬天氣乾燥，燥邪容易令氣管和上呼吸道感到乾燥，所以一般秋天容易出現咳嗽、感冒和鼻敏感等。補肺宜潤，所以秋燥必先潤肺。

食療方面：

銀耳雪梨湯（二人份量）

材料：乾銀耳1兩，雪梨2個，
　　　南杏4錢，無花果3個，
　　　瘦豬肉半斤，鹽適量。

做法：銀耳先浸清水約半小時，浸發後撕去硬蒂，撕細備用；雪梨去皮、去芯，切粗件；無花果在底部切十字花；瘦豬肉、南杏洗淨，與其餘材料放入適量清水同煲，水沸後轉慢火續煲 2.5 小時，最後下適量海鹽調味即可。

功效：銀耳生津潤肺、滋陰養胃，秋天時能滋潤乾燥的身體和皮膚，適合乾咳、便秘、皮膚乾燥的人士。

禁忌：咳嗽痰多、陽虛怕冷的人不宜吃銀耳。

冬天

寒主收引，故冬天多見咳嗽、氣管敏感和哮喘。冬天補肺，重在溫陽潤燥。

食療方面：

蟲草花紅蘿蔔淮山湯（二人份量）

乾蟲草花

材料：乾蟲草花 3/4 兩，紅蘿蔔（中）2 條，淮山 3/4 兩，豬肉半斤，蜜棗 2 粒，生薑 3 片。

做法：紅蘿蔔去皮，切粗件，與其餘材料一同清洗乾淨，加入適量清水同煲，水沸後轉慢火續煲 3 小時，最後下適量海鹽調味即可。

功效：補肺益腎、潤燥。

宜忌：孕婦、大病、久病和長期病患者，服用前先諮詢合資格的註冊中醫師。

肺臟與白色食品

在五臟中的「肺」，剛好對應五色中的「白色」，白色食物應該選擇能化痰、滋陰補氣的食物，例如枇杷、蜂蜜、蓮子、百合、杏仁、蓮藕、白芝麻等。

百合

具養陰、潤肺止咳、清心安神的作用，因其藥性平和，正是滋養肺陰的首選藥材。用其作為藥膳，可改善氣候帶來的影響。常用量為 1/4 兩至 4 錢。

百合煲雪梨（二人份量）

功效：滋陰潤肺、寧心止咳

材料：乾百合 4 錢，雪梨 1 個。

做法：百合洗淨，以半碗水浸泡一夜。翌日將百合連水一起放入鍋內，加半碗水，水滾後以中火至慢火煲 1.5 小時。將雪梨去皮、去芯，切成小塊，加入適量冰糖，與百合煮半小時。

服法：可早晚飲用。

麥冬

性甘寒質潤，有滋陰的功效，既能善於清養肺胃之陰，又可清心經之熱，是一味滋清兼備的補益良藥。臨床上常用麥冬配伍人參、五味子以增強滋陰潤燥的功效，而且可用以治療因肺陰不足而引起的喉嚨痕癢、咳嗽無痰、口渴咽乾及腸燥便秘等症。

天冬

味甘、苦，性寒，入肺、腎二經。用於肺陰虛的燥咳，腎陰虛的潮熱盜汗。有良好的抗腫瘤作用。

二冬粥

材料：天冬 1/4 兩，麥冬 3 錢，白米 3/4 兩。

做法：麥冬、天冬及水 1 公升放入煲內，水滾後煲 1 小時，隔渣，加入白米和開水煮成粥。

適用：養陰補津，作為糖尿病人的輔助食療。

沙參麥冬粥（三至四人份量）

沙參

材料：沙參、麥冬各 4 錢，白米 1 1/4 兩，海鹽適量。

做法：沙參、麥冬與水 1 公升煮沸，隔渣取汁。將汁液加入白米煮成粥，待粥將熟時，加入海鹽調味即可。

服法：每日食用一碗，有益氣養陰、潤肺生津、化痰止咳的功效。

天冬麥冬太子參茶

材料：天冬 2 錢，麥冬 3 錢，天花粉 1 錢，黃芩
1 錢，知母 1 錢，荷葉 1 錢，太子參 5 錢，
甘草 5 錢。

做法：所有藥材洗淨，加入適量水，水滾後煲 45
分鐘至 1 小時，隔渣，即可飲用。

服法：每天一碗。

功效：養陰潤肺、生津止渴，治理消渴（糖尿病）。

玉竹

味甘，性平，入肺、胃二經。有滋陰潤燥、除煩止咳
之功效。研究顯示，玉竹可通過提高胰島素的敏感性達到
輔助降低血糖的作用，特別適合糖尿病合併秋燥者食用。
常用量為 1/4 兩至 4 錢。

西洋參（花旗參）

味甘，微苦，性寒，
入心、肺、腎經，有補
肺降火、養胃生津之功。
不過，服用西洋參要注
意以下三點：

- 服用時不宜飲茶，因茶內有鞣酸，與西洋參的有效成
  分結合使吸收率下降；

- 服用後不宜吃白蘿蔔，因白蘿蔔是破氣的，西洋參則
  補氣；

- 畏寒肢冷、腹瀉、胃有寒濕、舌苔膩濁者不宜選用。

九款增強肺臟的食材

在中醫層面，增強肺臟功能的食物有很多，現舉例如下：

枇杷

中醫認為枇杷果實有潤肺、止咳、止渴的功效。因其當中含有胡蘿蔔素，並在水果界中居當首三位。在藥用方面，枇杷葉為常見中藥，其性味苦，微寒，歸肺、胃經，有清肺止咳、降逆止嘔的作用。

杏仁

中藥的杏仁分為甜與苦，甜杏仁常用於食療，有止咳效果；苦杏仁則多入藥，其毒性比一般甜杏仁略高，但經過煮熟後食用可降低毒性，可以與雪梨、雪耳燉煮，對秋冬劇烈咳嗽很有效。

雪梨

自古以來，雪梨被稱為「百果之宗」，根據中醫藥典，雪梨有潤肺、止咳、消痰降火的功效，特別適用於秋燥時，因內熱導致煩渴、咳喘等症。可是，雪梨畢竟屬於寒涼水果，因此體質虛寒、容易手腳冰冷、月經不順或長期咳嗽者，建議先把雪梨蒸熱、略

煮為佳。做法相對簡單，將雪梨內部掏空，放入川貝、冰糖入鍋煮透，待稍涼後加入蜜糖即可。不過，腸胃功能不佳、有長期腹瀉問題人士要注意，不宜多吃雪梨，以免腹瀉加劇。

百合

百合性微寒，入心、肺經。有清肺解毒的作用，是秋季養肺不可少的食療聖品。而在草藥運用上，經常用於潤肺、止咳、寧心安神之用。百合對有慢性咳嗽，尤其是乾咳的人效果特別顯著，因其含有黏液，可以滋潤肺臟，同時有去熱的作用。

銀耳（雪耳）

銀耳被譽為「長生不老藥」，有「菌中之王」的美稱，味甘淡、性微寒，入肺、胃、腎經，具生津潤肺、益氣活血、滋陰養胃、補腦強心的作用。針對便秘、腸胃燥熱者也可使用。但要注意，銀耳偏寒，受風寒引起的感冒，出現怕冷、咳嗽、痰多清稀如水者忌食。

海底椰

海底椰為夏天常見的湯料之一，有滋陰潤肺、除煩清熱、潤肺止咳等作用。海底椰果肉細白，美味可口，滋陰壯陽，還能治療中風、精神煩躁等症。

川貝

川貝雖有止咳功效，可用於煲湯，但要留意兼有清熱的功效，對熱咳患者會達到效果。相反，寒咳患者則不宜吃川貝，否則將有機會加劇咳嗽。

沙參

味甘，性微寒，歸肺、胃經。有養陰潤肺、益生津之功，適用於陰虛肺燥或熱傷肺陰所致的乾咳痰少。

款冬花

味辛，性溫，入肺經，有潤肺下氣、止咳化痰之功。款冬花常經蜜炙之後使用，稱為炙冬花，有增強潤肺止咳、平喘的作用。臨床上，冬花多與其他藥物配伍以增強療效，如冬花重在止咳，紫菀重在祛痰，止咳方中，二藥常配伍使用，共奏化痰止咳之效。若與麻黃、杏仁、蘇子為伍，稱款冬定喘湯，治痰嗽哮喘遇冷即發之症，療效頗佳。

滋潤湯水

Healthy soups

為家人送上潤肺強身的家常湯，
增強抵抗力，讓身體加加油！

Starfruit, Dried Ya-li Pear and Pork Shank Soup

楊桃雪梨乾豬腱湯

● 清熱化痰、降火潤肺 ●

材料

楊桃 .... 2 個
雪梨乾 .... 2 兩
豬腱 .... 8 兩
蜜棗 .... 2 粒
鹽 .... 半茶匙

做法

❶ 豬腱洗淨，切大塊，放入滾水中飛水。

❷ 楊桃洗淨，切成星狀厚片；雪梨乾、蜜棗洗淨。

❸ 湯煲內注入清水 10 杯煲滾，放入豬腱、雪梨乾、楊桃、蜜棗煲滾 10 分鐘，轉小火煲 2 小時，下鹽調味即成。

小貼士

楊桃清熱止渴、化痰止咳、潤肺順氣，對風熱咳嗽及咽喉疼痛有改善作用；雪梨乾潤燥生津、清熱化痰，兩者合用適合秋天煲湯飲用。

STARFRUIT, DRIED YA-LI PEAR AND PORK SHANK SOUP

clearing Heat; dissolving phlegm; reducing excess body heat; nourishing the Lungs

Ingredients

2 starfruits
75 g dried Ya-li pear
300 g pork shank
2 candied dates
1/2 tsp salt

Method

1. Rinse and cut pork shank into big pieces. Scald with boiling water.
2. Rinse starfruits and cut into star-shaped thick slices. Rinse dried Ya-li pear and candied dates.
3. Pour 10 cups of water into pot, Bring to boil. Add pork shank, dried Ya-li pear, starfruits and candied dates. Bring to boil for 10 minutes. Adjust to low heat and simmer for 2 hours. Add salt and serve.

TIPS

Starfruit can help clear Heat, quench thirst, dissolve phlegm, relieve cough, nourish the Lungs and improve breathing, great for treating Wind-Heat cough and sore throat. Dried Ya-li pear is good for moistening dryness, stimulating saliva secretion, clearing Heat, dissolving phlegm. These two ingredients are good for making healthy soup for the autumn.

White Crucian Carp Soup with Fresh Lily Bulbs and Papaya

鮮百合木瓜魚湯

● 潤肺養顏、補中益氣 ●

材料

鮮百合 .... 2 球
半生熟木瓜 .... 1 斤
白鯽魚 .... 1 條（約 12 兩）
瘦肉 .... 4 兩
薑 .... 2 片

做法

❶ 鮮百合切去頭尾兩端及焦黑部分，撕成瓣狀，洗淨備用。

❷ 木瓜開邊，用小匙刮去籽，每邊切成三大塊，切皮，洗淨。

❸ 瘦肉洗淨，切厚片。

❹ 白鯽魚劏好，洗淨，抹乾水分。放入油鑊內，加入薑片煎至魚兩面呈微黃色，隔去多餘油分，傾入滾水 10 杯，用大火煮 10 分鐘，加入瘦肉及木瓜煮滾，轉中火煲半小時，最後下鮮百合煲 10 分鐘，下少許鹽調味即成。

小貼士

鮮百合具潤肺止咳、養顏安神
之效。用新鮮百合煲湯,湯水
清甜美味,但鮮品不耐存。

WHITE CRUCIAN CARP SOUP WITH FRESH LILY BULBS AND PAPAYA

nourishing the Lungs; improving the complexion; enriching Qi; nourishing the body

Ingredients

2 fresh lily bulbs
600 g half-ripe papaya
1 white crucian carp (450 g)
150 g lean pork
2 slices ginger

TIPS

Fresh lily bulbs have the function of nourishing the Lungs, stopping cough, improving skin and calming the nerves. Soups with fresh lily bulbs are sweet and delicate but they cannot last long.

Method

1. Cut off the head, root and the black parts from lily bulbs. Tear into small pieces. Rinse and set aside.
2. Cut papaya into halves. Remove seeds with a teaspoon. Cut each half into 3 large pieces. Cut off the skin and rinse.
3. Rinse lean pork and cut into thick slices.
4. Cut open and gut white crucian carp, rinse and wipe dry. Fry white crucian carp with ginger slices and oil until both sides turn light brown. Drain the oil. Pour 10 cups of boiling water and boil over high heat for 10 minutes. Put in lean pork and papaya and bring to boil. Turn to medium heat and boil for 30 minutes. Put in lily bulbs and boil for 10 minutes. Season with salt. Serve.

養好肺 • 強體質

23

Crocodile Meat Soup with Fig and Luo Han Guo

無花果羅漢果燉 鱷魚肉湯

● 清熱化痰、潤肺止咳 ●

clearing Heat; expelling phlegm; nourishing the Lungs;
stopping cough

材料

鱷魚肉.......... 450 克（12 兩）
豬骨 ....半斤
無花果........................ 1 個
羅漢果........................半個
薑................................數片
水................................ 5 杯

調味料

鹽.................................適量

做法

❶ 鱷魚肉和豬骨一同飛水，
　沖淨。
❷ 無花果、羅漢果洗淨。
❸ 全部材料先用大火煲滾，
　然後轉放入燉盅內，隔水
　燉 2 小時，加鹽調味。

Ingredients

450 g dried crocodile meat
300 g pork bones
1 dried fig
1/2 Luo Han Guo
ginger slices
5 cups water

Seasoning

salt

Method

1. Scald crocodile meat and pork bones. Rinse.
2. Rinse fig and Luo Han Guo.
3. Bring all ingredients to boil over high heat. Transfer in a stewing pot and stew for 2 hours. Season with salt. Serve.

小貼士 / TIPS

羅漢果化痰止咳、益肝潤肺。宜選個
體圓渾、色黃褐及輕搖不響的。

Luo Han Guo can expel phlegm；stop
cough; benefit the Liver and nourish the
Lungs. Choose the one which has round
shape with cinnamon colour and no
sound when jiggling.

養好肺 • 強體質

Pork Shoulder Soup with Wu Zhi Mao Tao, Sha Shen and Yu Zhu

五指毛桃沙參 玉竹西施骨湯

● 滋潤化痰、舒筋活絡 ●

dissolving phlegm; soothing sinew and tendon; activating meridian

材料

五指毛桃 .... 3 兩
沙參、玉竹................. 各 1 兩
西施骨........................ 12 兩
甘筍 .... 8 兩
蜜棗 .... 2 粒
薑............................... 2 片
鹽............................. 半茶匙

做法

❶ 西施骨洗淨，放入煲內加入浸過面的清水，用大火煲至大滾，取出西施骨洗淨，備用。

❷ 甘筍去皮，洗淨，切塊。

❸ 五指毛桃、沙參、玉竹、蜜棗、薑片全部洗淨，放入湯煲，注入清水 12 碗煲滾，加入西施骨及甘筍煲滾 10 分鐘，轉小火煲 2 小時，下鹽調味即可。

Ingredients

113 g Wu Zhi Mao Tao
38 g each of Sha Shen, Yu Zhu
450 g pork shoulder blade
300 g carrot
2 candied dates
2 slices ginger
1/2 tsp salt

Method

1. Rinse pork bone. Put in a pot and add water to cover the bones. Bring to vigorous boil with high heat. Dish up pork bone and rinse well.
2. Peel, rinse and cut carrot into pieces.
3. Rinse Wu Zhi Mao Tao, Sha Shen, Yu Zhu, candied dates and ginger. Put all into soup pot. Add 12 bowls of water. Bring to boil. Add pork bone and carrot. Boil for 10 minutes. Adjust to low heat and simmer for 2 hours. Add salt and serve.

小貼士 / TIPS

五指毛桃補脾益氣、祛痰補肺，煲出來的湯水帶陣陣椰子香氣，適合秋天滋潤喉嚨及肺臟，新鮮的可於山草藥店購買。
Wu Zhi Mao Tao can nourish the Spleen, enrich Qi, dispel phlegm and nourish the Lungs. The soup will have a coconut fragrance and is good for soothing throat and the Lungs during autumn. The fresh one is available in shops selling wild herbs.

養好肺 • 強體質

Pork Shin Soup with Chuan Bei, White Fungus and Almonds

川貝雪耳杏仁 豬腱湯

● 滋陰潤肺、化痰止咳 ●

nourishing the Yin; moistening the Lungs;
expelling phlegm; stopping cough

小貼士 / TIPS

川貝味甘苦、性微寒，有潤肺止咳、清熱化痰之效。
Chuan Bei tastes bittersweet and is mildly Cold in nature. It nourishes the Lungs and stops cough, clears Heat and expels phlegm.

材料

川貝 ....半兩
雪耳 ....3/4 兩
南北杏.......................... 1 兩
豬腱 .... 12 兩
無花果......................... 3 個
陳皮 .... 1/3 個

做法

❶ 雪耳用水浸軟（約 1 小時），剪去硬蒂，摘成小朵，洗淨，飛水，隔去水分。

❷ 陳皮用水浸軟，刮淨瓤；川貝及南北杏洗淨。

❸ 豬腱洗淨，切大塊，飛水，過冷河，瀝乾水分。

❹ 煲內注入清水 12 杯，加入川貝、無花果、南北杏及陳皮煲滾，下豬腱煲 15 分鐘，轉小火煲 1 小時，加入雪耳再煲半小時，下少量鹽調味，雪耳伴湯食用。

Ingredients

19 g Chuan Bei
28 g white fungus
38 g bitter and sweet almonds
450 g pork shin
3 dried figs
1/3 dried tangerine peel

Method

1. Soak white fungus for about 1 hour until soft. Cut off the hard stems and tear into small pieces. Rinse, scald and drain.

2. Soak dried tangerine peel until soft. Scrape off the pith. Rinse Chuan Bei and almonds.

3. Rinse pork shin and cut into large pieces. Scald, rinse and drain.

4. Pour 12 cups of water in a pot. Put in Chuan Bei, dried figs, almonds and dried tangerine peel. Bring to boil. Put in pork shin and boil for 15 minutes. Turn to low heat and simmer for 1 hour. Put in white fungus and boil for 30 minutes. Season with salt. Serve with white fungus.

Chayote and Fig Soup with Almonds and Lean Pork

合掌瓜無花果
杏仁瘦肉湯

● 清熱健脾、潤肺化痰、增強免疫力 ●

Clearing Heat; strengthening the Spleen; nourishing the Lungs;
dissolving phlegm; improving immunity

小貼士 / TIPS

無花果切半煲湯，湯水略帶酸味；如長時間煲湯，建議原個使用。
Soup made with dried fig that is already cut into half will have a slight
sour taste. For long simmering soup, it is better to use the whole fig.

材料

合掌瓜............................ 1 個
無花果............................ 4 個
南杏 .... 1 兩
北杏 ....2 茶匙
陳皮 .... 1/3 個
瘦肉 .... 8 兩

做法

❶ 陳皮用水浸 1 小時，刮淨
　內瓤，洗淨。

❷ 瘦肉洗淨，切塊；合掌瓜
　洗淨，開邊去核，切大塊。

❸ 無花果、南北杏一同洗淨。

❹ 合掌瓜、無花果、南北杏、
　陳皮放入煲內，注入水 10
　碗煲滾，加入瘦肉再煲，
　轉中火煲 15 分鐘，再轉小
　火煲 1.5 小時，一併連湯
　料食用。

Ingredients

1 chayote
4 dried figs
38 g sweet almonds
2 tsp bitter almonds
1/3 dried tangerine peel
300 g lean pork

Method

1. Soak dried tangerine peel in water
   for 1 hour. Scrape off the pith. Rinse.
2. Rinse and cut lean pork into pieces.
   Rinse chayote, cut into half. Remove
   the seeds and cut chayote into big
   pieces.
3. Rinse dried figs and the almonds.
4. Put chayote, dried figs, almonds and
   dried tangerine peel in pot. Add 10
   bowls of water. Bring to boil. Add
   pork. Bring to boil. Adjust to medium
   heat and cook for 15 minutes.
   Adjust to low heat and simmer for
   1.5 hours. Serve the soup and the
   ingredients.

Tai Zi Shen Soup with Mai Dong and Lean Pork

太子參麥冬瘦肉湯

● 生津潤肺、清心安神 ●

improving saliva secretion; moistening the Lungs;
clearing thoughts; calming emotion

材料

太子參.......25 克（約 2/3 兩）
麥冬 .... 1 兩
瘦肉 .... 12 兩
紅棗 .... 4 粒

做法

❶ 瘦肉洗淨，切薄片。

❷ 紅棗去核，用刀拍鬆，洗
淨。

❸ 太子參、麥冬一同洗淨。

❹ 太子參、麥冬及紅棗放入
煲內，加入清水 12 碗煲
滾，下瘦肉煲滾，轉小火
再煲 1 小時，待暖飲用。

Ingredients

25 g Tai Zi Shen
38 g Mai Dong
450 g lean pork
4 red dates

Method

1. Rinse and cut lean pork into pieces.
2. Remove the stone of red date. Smash and rinse well.
3. Rinse Tai Zi Shen and Mai Dong.
4. Add Tai Zi Shen, Mai Dong and red date to pot. Add 12 bowls of water. Bring to boil. Add lean pork. Bring to boil. Adjust to low heat and simmer for 1 hour. Serve warm.

小貼士 / TIPS

太子參補益脾肺，對心悸不眠、精神疲憊有一定幫助。
Tai Zi Shen can help nourish the Spleen and Lungs which is good for alleviating palpitation led insomnia and tiredness.

Partridge Soup with Papaya and Coco-de-Mer

木瓜海底椰
杏仁鷓鴣湯

● 健肺止咳 ●

strengthening the Lungs; stopping cough

小貼士 / TIPS

鷓鴣煲湯能健肺、化痰、止咳，具保健療效。
Partridge strengthens the Lungs, stops cough and expels phlegm.

材料

木瓜 .... 1 個
　　　（半生熟，重約 12 兩）
乾海底椰 .... 1 兩
冰鮮鷓鴣 .... 1 隻
瘦肉 .... 4 兩
南杏 .... 1 兩
陳皮 .... 1/3 個

做法

❶ 陳皮用水浸軟，刮去瓤。

❷ 鷓鴣處理乾淨，洗淨，與
　 瘦肉同飛水，過冷河，瀝
　 乾水分。

❸ 木瓜去皮、去籽，洗淨，
　 切塊。

❹ 海底椰、南杏洗淨。

❺ 燒滾清水 15 杯，放入鷓
　 鴣、瘦肉、海底椰、南杏
　 及陳皮，用大火煲 20 分
　 鐘，轉小火煲 1.5 小時，
　 最後加入木瓜再煲 1 小時，
　 下鹽調味。

Ingredients

1 papaya (about 450 g, half-ripe)

38 g dried coco-de-mer

1 chilled partridge

150 g lean pork

38 g sweet almonds

1/3 dried tangerine peel

Method

1. Soak dried tangerine peel until soft. Scrape off the pith.

2. Gut and dress the partridge. Rinse well. Scald with lean pork. Rinse and drain.

3. Peel the papaya, remove the seeds. Rinse and cut into pieces.

4. Rinse dried coco-de-mer and sweet almonds.

5. Bring 15 cups of water to the boil. Put in partridge, lean pork, dried coco-de-mer, sweet almonds and dried tangerine peel. Boil over high heat for 20 minutes. Turn to low heat and simmer for 1.5 hours. Put in papaya at last and simmer for 1 hour. Season with salt and serve.

Huai Shan, Fox Nuts and Flathead Fish Soup

淮山芡實牛鰍魚湯

● 健脾益肺，增強體質 ●

strengthening the Spleen; benefiting the Lungs; improving the body

材料

牛鰍魚.......................... 12 兩
板豆腐.......................... 2 塊
瘦肉 .... 4 兩
淮山、芡實................ 各 1 兩
薑 .... 3 片

做法

❶ 淮山用水浸半小時，洗淨。

❷ 瘦肉洗淨，飛水，瀝乾水分。豆腐、芡實洗淨。

❸ 牛鰍魚處理乾淨，洗淨，抹乾水分。

❹ 燒熱油鑊，下薑片及牛鰍魚煎至兩面微黃，隔去油分。

❺ 燒滾清水 10 杯，放入全部材料，用大火煲 20 分鐘，轉小火煲 1.5 小時，下鹽調味。

Ingredients

450 g flathead fish

2 pieces hard bean curd

150 g lean pork

38 g Huai Shan

38 g fox nuts

3 slices ginger

Method

1. Soak Huai Shan in water for half an hour. Rinse.

2. Rinse and scald lean pork. Drain well. Rinse bean curd and fox nuts.

3. Gut and rinse flathead fish. Wipe dry.

4. Heat the oil. Fry flathead fish with ginger until light brown. Drain.

5. Bring 10 cups of water to the boil. Put in all ingredients and boil over high heat for 20 minutes. Turn to low heat and simmer for 1.5 hours. Season with salt and serve.

小貼士 / TIPS

淮山健脾補氣、益肺滋腎；芡實補脾益腎，兩者用來煲湯煮粥，對身體有保健作用。
Huai Shan benefits the Spleen, promotes Qi, tones the Lungs and nourishes the Kidneys. Fox nuts tones the Spleen and benefits the Kidneys. They are commonly used in soups and congee. They are highly beneficial to the body.

Fish Maw Soup with Monkey Head Mushrooms

猴頭菇滋潤湯

● 健脾潤肺、除煩止咳 ●

benefiting the Spleen; nourishing the Lungs;
calming the nervous; stopping cough

小貼士 / TIPS

猴頭菇是近年被廣泛應用的煲湯材料，能提高抗病能力。煮前先浸軟及擠乾水分。

Monkey head mushroom becomes a popular soup ingredient since last few years. It can improve immunity. Make sure to soak until soft and squeeze the water before use.

材料

花膠 .... 1.5 兩
猴頭菇 .... 3/4 兩
淮山 .... 3/4 兩
玉竹 .... 1/4 兩
沙參 .... 3/4 兩
桂圓肉 .... 半兩
杞子 .... 2 湯匙
豬骨 .... 1 斤
水 .... 8 杯

調味料

鹽 .... 適量

做法

❶ 煲湯前一晚，燒滾 3 杯水，放入薑葱煮滾，放入花膠焗一夜，至浸發透後洗淨，切件。

❷ 猴頭菇、淮山、玉竹、沙參、桂圓肉和杞子沖淨。

❸ 豬骨飛水，洗淨。

❹ 用大火煲滾所有材料，轉用中慢火煲約 2 小時，加入適量鹽調味即可飲用。

Ingredients

57 g fish maw
30 g monkey head mushrooms
30 g Huai Shan
10 g Yu Zhu
30 g Sha Shen
19 g dried longan
2 tbsp Qi Zi
600 g pork bones
8 cups water

Seasoning

salt

Method

1. On the night before, bring 3 cups of water to boil and add ginger and spring onion. Add fish maw and remain covered overnight. Rinse and cut into pieces.

2. Rinse monkey head mushrooms, Huai Shan, Yu Zhu, Sha Shen, dried longan, Qi Zi.

3. Scald pork bones and rinse.

4. Bring all ingredients to boil over high heat. Turn to medium low heat and boil for about 2 hours. Season with salt. Serve.

養好肺 • 強體質

Pork Lung Soup with Mulberry Leaf and Almond Juice

桑葉杏汁豬肺湯

● 止咳平喘、清肺熱 ●

材料

乾桑葉............................半兩
南北杏.........................1.5 兩
豬肺 ....1 個
瘦肉 ....半斤
無花果..........................2 個
薑.................................2 片

做法

❶ 豬肺灌水洗淨（此步驟可請肉販代勞），切塊，洗淨。豬肺放入白鑊內，煎片刻至轉成白色，盛起，過冷河，洗淨，瀝乾水分。

❷ 南北杏洗淨，用水浸 1 小時，隔去水分。南北杏放於攪拌機內，加水 1.5 杯，磨成幼滑的杏仁漿，用煲魚袋過濾，冷藏備用。

❸ 瘦肉洗淨，切片；乾桑葉洗淨。

❹ 煲滾清水 8 杯，下瘦肉、豬肺、乾桑葉、無花果及薑片，用大火煲 15 分鐘，轉小火煲 1.5 小時（豬肺軟腍即可），去掉桑葉、薑片及瘦肉，傾入杏汁拌勻，用小火煮滾片刻，下少許鹽調味即可，豬肺可伴湯食用。

小貼士

到肉檔必須購買已灌水洗淨、呈淡粉紅色及無血水的豬肺。

PORK LUNG SOUP WITH MULBERRY LEAF AND ALMOND JUICE

stopping cough; alleviating difficulty in breathing; clearing Heat in the Lungs

Ingredients

19 g dried mulberry leaves
57 g bitter and sweet almonds
1 pork lung
300 g lean pork
2 dried figs
2 slices ginger

TIPS

You should always buy clean, washed pork lung with pale pink colour and no blood from butchers.

Method

1. Wash pork lung by piping water through it (you may ask the butcher to do it.). Cut into pieces and rinse. Fry pork lung in a dry wok until it turns white. Remove and rinse with cold water. Drain.
2. Rinse almonds, soak for 1 hour and drain. Put the almonds in a blender. Pour in 1.5 cups of water and blend into almond juice. Sift with a cloth bag. Refrigerate.
3. Rinse lean pork and slice; rinse dried mulberry leaves.
4. Bring 8 cups of water to boil. Put in lean pork, pork lung, mulberry leaves, dried figs and ginger. Boil over high heat for 15 minutes. Turn to low heat and simmer for 1.5 hours, or until pork lung softens. Discard mulberry leaves, ginger and lean pork. Pour in almond juice and mix well. Boil over low heat. Season with salt. Serve with pork lung.

Teal Soup with Tai Zi Shen and Huai Shan

太子參淮山水鴨瘦肉湯

● 益肺健脾，強身健體 ●

材料

冰鮮水鴨 .... 1 隻
瘦肉 .... 6 兩
太子參 .... 半兩
淮山 .... 1 兩
無花果 .... 4 個
陳皮 .... 2/3 個

做法

❶ 陳皮用水浸 1 小時，刮淨內瓤，洗淨。

❷ 淮山洗淨，用水浸 1 小時，隔去水分。

❸ 太子參、無花果洗淨，瀝乾水分；瘦肉洗淨，切塊。

❹ 水鴨洗淨，放入滾水內飛水 3 分鐘，取出洗淨，瀝乾水分。

❺ 淮山、太子參、無花果、陳皮放入煲內，注入水 10 碗煲滾，放入水鴨、瘦肉煲滾，轉中火煲 15 分鐘，轉小火再煲 2 小時即成。

小貼士

太子參補氣養陰，與淮山、無花果煲成湯，有滋潤的功效。

TEAL SOUP WITH TAI ZI SHEN AND HUAI SHAN

Enriching the Lungs; strengthening the Spleen; improving the health

Ingredients

1 chilled teal
225 g lean pork
19 g Tai Zi Shen
38 g Huai Shan
4 dried figs
2/3 dried tangerine peel

Method

1. Soak dried tangerine peel in water for 1 hour. Scrape off the pith and rinse.
2. Rinse Huai Shan, soak in water for 1 hour, drain.
3. Rinse Tai Zi Shen and dried figs, drain. Rinse and cut lean pork into pieces.
4. Rinse teal, scald in boiling water for 3 minutes. Remove teal and rinse well, drain.
5. Add Huai Shan, Tai Zi Shen, dried figs and dried tangerine peel to pot. Add 10 bowls of water. Bring to boil. Add teal and lean pork. Bring to boil. Adjust to medium heat and cook for 15 minutes. Adjust to low heat and simmer for 2 hours. Serve.

TIPS

Tai Zi Shen replenishes Qi and nourishes the Yin. Huai Shan can help strengthen the Spleen and replenish body deficencies. Combing Huai Shan with dried fig, it can make a health nourishing soup.

養好肺 • 強體質

45

Pork Soup with Yacon

天山雪蓮果老火湯

● 清肝解毒、潤肺祛火 ●

purifying the Liver; expelling toxins; nourishing the Lungs and Heat

材料

豬骨	12 兩
天山雪蓮果	半斤
雪耳	1 兩
雪梨	1 個
蓮子	1 兩
生熟薏米	1 兩
水	8 杯

調味料

鹽 ....適量

做法

❶ 豬骨洗淨後飛水。

❷ 天山雪蓮果、雪梨去皮，切件。

❸ 雪耳用水浸軟去蒂，分拆小朵。

❹ 蓮子去芯，與生熟薏米一同洗淨。

❺ 用大火煲滾所有材料，再轉用中慢火煲約 2 小時，可以加入適量鹽調味。

Ingredients

450 g pork bones

300 g yacon

38 g white fungus

1 Ya-li pear

38 g lotus seeds

38 g raw and cooked job's tears

8 cups water

Seasoning

salt

Method

1. Rinse and scald pork bones.
2. Peel yacon and Ya-li pear. Cut into pieces.
3. Soak white fungus until soft. Tear off hard stems and tear into small pieces.
4. Core lotus seeds and rinse with job's tears.
5. Bring all ingredients to boil over high heat. Turn to medium-low heat and boil for 2 hours. Season with salt. Serve.

天山雪蓮清潤多汁，有滋養益氣、清肝解毒之效，適合秋冬煲湯飲用。以身重及外皮完整的為佳。

With its juicy, yacon has the function of nourishing Qi and the Liver, expelling toxins. It is suitable for making soup in autumn and winter. Choose heavy yacon with unblemished skin.

養好肺 • 強體質

Filefish Soup with Jinhua Ham, Sha Shen and Yu Zhu

沙參玉竹
金腿沙鯭湯

● 清潤肺部 ●

clearing and moistening the Lungs

材料

沙鯭魚.................. 1 斤（大）
瘦肉 .... 4 兩
金華火腿 .... 1 兩
沙參 .... 1/3 兩
玉竹 ....半兩
薑.................................. 3 片

Ingredients

600 g filefish (big-sized)
150 g lean pork
38 g Jinhua ham
12 g Sha Shen
19 g Yu Zhu
3 slices ginger

做法

❶ 瘦肉洗淨，飛水，瀝乾水分。

❷ 沙參、玉竹、金華火腿及薑洗淨。

❸ 沙鯭魚洗淨，抹乾水分。

❹ 燒熱油鑊，下薑片及沙鯭魚煎至兩面微黃，隔去油分。

❺ 燒滾清水 10 杯，放入全部材料煲 15 分鐘，轉小火煲 2 小時，下鹽調味即成。

Method

1. Rinse lean pork. Scald and drain.
2. Rinse Sha Shen, Yu Zhu, Jinhua ham and ginger.
3. Rinse filefish and wipe dry with kitchen paper.
4. Heat the oil. Fry filefish with ginger until filefish becomes light brown. Drain.
5. Bring 10 cups of water to the boil. Put in all ingredients and boil over high heat for 15 minutes. Turn to low heat and simmer for 2 hours. Season with salt and serve.

小貼士 / TIPS

大沙鯭魚骨頭大、魚肉厚,魚味濃郁,適合長時間煲煮。

Big filefish has thicker fleshes and bigger skeleton. It has stronger taste and is suitable for long-boiled soup.

Beef Shin Soup with Dried Tangerine Peel, Lily Bulb and Lotus Root

陳皮百合牛腱蓮藕湯

● 潤肺下氣（降氣）、補虛養腎 ●

nourishing the Lungs; calming the adverse-rising energy;
replenishing the weak body and the Kidneys

小貼士 / TIPS

原條牛腱煲湯，可保存牛腱之原味，以免肉質粗韌；蓮藕切件煲湯，
其澱粉質會影響湯水之清澈。

Cooking soup with a whole of beef shin, it can preserve the taste of
beef shin and it would not be hard and tough. If you boil the soup with
lotus root sliced, its starch would make the soup not clear.

材料

牛腱 .... 12 兩
蓮藕 .... 1 斤
乾百合 .... 1.5 兩
陳皮 ....半個

做法

❶ 陳皮用水浸軟，刮淨瓤。

❷ 百合洗淨，備用。

❸ 蓮藕洗淨污泥，切掉蓮藕節。

❹ 牛腱洗淨，飛水，過冷河，瀝乾水分。

❺ 煲內注入清水 14 杯，下陳皮煲滾，加入牛腱及蓮藕煲 15 分鐘，轉小火煲 1 小時，最後加入百合再煲 1 小時，取出牛腱及蓮藕切片，放回湯內煲滾，下少許鹽調味，飲湯時可連材料一併享用。

Ingredients

450 g beef shin
600 g lotus root
57 g dried lily bulbs
1/2 dried tangerine peel

Method

1. Soak dried tangerine peel until soft. Scrape off the pith.
2. Rinse lily bulbs and set aside.
3. Rinse off the mud thoroughly from lotus root. Cut off the joint between the lotus root.
4. Rinse beef shin. Scald, rinse again and drain.
5. Pour in 14 cups of water in a pot. Put in dried tangerine peel and bring to boil. Put in beef shin and lotus root and boil for 15 minutes. Turn to low heat and simmer for 1 hour. Put in lily bulbs and simmer for 1 hour. Slice beef shin and lotus root and put back in a pot. Bring to boil. Season with salt. Serve with the ingredients.

Qing Bu Liang with Pigeon and White Fungus

雪耳清補涼乳鴿湯

● 滋陰、潤肺、補腎 ●

nourishing the Yin; moistening the Lungs;
strengthening the Kidneys

小貼士 / TIPS

有少許感冒症狀者，不適合飲用此湯。
This soup is not suitable for those who have symptoms of influenza.

材料

乳鴿 .... 1 隻
瘦肉 .... 6 兩
雪耳 ....3/4 兩
清補涼............1 包（約 3 兩）
蜜棗 .... 2 粒
陳皮 ....半個

做法

❶ 陳皮用水浸軟，刮淨內瓤。

❷ 雪耳用水浸軟，剪去硬蒂，
洗淨，飛水，過冷河。

❸ 乳鴿洗淨，飛水，過冷河。

❹ 瘦肉洗淨，切厚片。

❺ 清補涼及蜜棗同洗淨。

❻ 煲內注入清水 14 杯，加入
陳皮、清補涼及蜜棗煲滾，
放入乳鴿及瘦肉，用大火
煲 15 分鐘，轉小火煲 1
小時，加入雪耳煲半小時，
下少許鹽調味。

Ingredients

1 young pigeon
225 g lean pork
30 g white fungus
1 pack (113 g) Qing Bu Liang
2 candied dates
1/2 dried tangerine peel

Method

1. Soak dried tangerine peel until soft. Scrape off the pith.

2. Soak white fungus until soft. Cut off the hard stems and rinse. Scald and rinse again.

3. Rinse pigeon. Scald and then rinse.

4. Rinse lean pork and cut into thick slices.

5. Rinse Qing Bu Liang and candied dates.

6. Pour 14 cups of water in a pot. Put in dried tangerine peel, Qing Bu Liang and candied dates. Bring to boil. Put in pigeon and lean pork and boil over high heat for 15 minutes. Turn to low heat and simmer for 1 hour. Put in white fungus and boil for 30 minutes. Season with salt. Serve.

養好肺 ‧ 強體質

Apple, Pear and Coco-de-Mer Soup with Almonds and Pork

海底椰蘋果雪梨 杏仁瘦肉湯

● 鎮咳除痰，有助消化，紓緩感冒後咳嗽 ●

Calming cough; dissolving phlegm; improving digestion; relieving post flu cough

小貼士 / TIPS

海底椰止咳化痰、清燥熱，加上滋潤的蘋果、雪梨及杏仁，有潤肺強身之效。選購海底椰以乾品為佳。

Coco-de-mer has the ability to relieve cough, dissolve phlegm and clear Heat. Combine it with apple, Ya-li pear and almonds, this soup can moisten the Lungs and strengthen the health. Dried coco-de-mer is recommended.

材料

海底椰	1 兩
蘋果	2 個
鴨咀梨	2 個
南杏	1 兩
北杏	2 茶匙
陳皮	1/3 個
蜜棗	3 粒
瘦肉	8 兩

做法

❶ 陳皮用水浸 1 小時，刮淨內瓤，洗淨。

❷ 瘦肉洗淨，切厚片。

❸ 蘋果及鴨咀梨洗淨，各切成 4 件，去蒂及去籽。

❹ 海底椰、南北杏、蜜棗一同洗淨。

❺ 將海底椰、南北杏、蜜棗、陳皮、蘋果、鴨咀梨放在煲內，注入水 10 碗煲滾，下瘦肉煲滾，轉中火煲 15 分鐘，再轉小火煲 1.5 小時即可。

Ingredients

38 g coco-de-mer

2 apples

2 Ya-li pear

38 g sweet almonds

2 tsp bitter almonds

1/3 dried tangerine peel

3 candied dates

300 g lean pork

Method

1. Soak dried tangerine peel in water for 1 hour. Scrape off the pith and rinse.

2. Rinse and cut lean pork into thick slices.

3. Rinse apple and pear, cut into quarters, remove stem and seeds.

4. Rinse coco-de-mer, sweet and bitter almonds and candied dates.

5. Add coco-de-mer, almonds, candied dates, dried tangerine peel, apple and pear to the pot. Add 10 bowls of water. Bring to boil. Add lean pork. Bring to boil. Adjust to medium heat and cook for 15 minutes. Adjust to low heat and simmer for 1.5 hours. Serve.

Crucian Carp and Watercress Soup

鯽魚西洋菜湯

● 清熱潤肺、化痰止咳 ●

clearing Heat; nourishing the Lungs; dissolving phlegm; stopping cough

小貼士 / TIPS

白鯽魚呈銀灰色，骨多，適宜煲成老火湯，價錢便宜。
White crucian carp is silverfish and bony, it is suitable for making long-boiled soup.

材料

白鯽魚............1 條（約 1 斤）
西洋菜..........................1 斤
甘筍 ....半斤
陳皮 ....2/3 個
鹽 ....3/4 茶匙

做法

❶ 西洋菜用水浸半小時，清洗三次至乾淨，瀝乾水分。

❷ 陳皮用水浸半小時，刮淨內瓤，洗淨。

❸ 甘筍去皮，洗淨，切塊。

❹ 白鯽魚去鱗劏好，洗淨，抹乾水分，下鹽抹勻，放入油鑊煎至兩面微黃，備用。

❺ 清水 10 杯放入煲內，加入陳皮煲滾，下西洋菜、甘筍、鯽魚，用中火煲 15 分鐘，轉小火再煲 1.5 小時即成。

Ingredients

1 white crucian carp (about 600 g)
600 g watercress
300 g carrot
2/3 dried tangerine peel
3/4 tsp salt

Method

1. Soak watercress in water for 1/2 hour. Rinse three times to thoroughly clean. Drain watercress.
2. Soak dried tangerine peel in water for 1/2 hour. Scrape off the peel and rinse.
3. Peel, rinse and cut carrot into pieces.
4. Gut fish and remove scales. Rinse and pat dry fish. Rub with salt. Pan-fry in wok with oil until both sides of fish turn golden. Set aside.
5. Pour 10 cups of water into pot. Add dried tangerine peel. Bring to boil. Add watercress, carrot and fish. Simmer over medium heat for 15 minutes. Adjust to low heat and simmer for 1.5 hours. Serve.

養好肺 • 強體質

57

Stewed Quail Soup with Lily Bulb and Sha Shen

百合沙參燉鵪鶉湯

● 潤肺止咳、清熱養陰 ●

nourishing the Lungs; stopping cough;
clearing Heat; nourishing the Yin

材料

鵪鶉 .... 2 隻
豬腱 .... 半斤
乾百合 .... 1 兩
沙參 .... 1.5 兩
水 .... 4 杯

調味料

鹽 .... 適量

做法

❶ 鵪鶉飛水，洗淨；豬腱飛水，洗淨。

❷ 百合、沙參洗淨。

❸ 將所有材料煲滾，然後轉放入燉盅內燉 2 小時。

❹ 加入適量鹽調味，即可享用。

Ingredients

2 quails

300 g pork shin

38 g dried lily bulbs

57 g Sha Shen

4 cups water

Seasoning

salt

Method

1. Scald quails and rinse; scald pork shin and rinse.
2. Rinse lily bulbs and Sha Shen.
3. Bring all ingredients to boil. Transfer to a stewing pot and stew for 2 hours.
4. Season with salt. Serve.

養好肺 • 強體質

Lotus Root and Dried Octopus Soup with Red Bean and Pork Bone

蓮藕章魚紅豆西施骨湯

● 補血健脾、補肺腎 ●

Replenishing Blood; strengthening the Spleen; nourishing the Lungs and Kidneys

小貼士 / TIPS

蓮藕、章魚及紅豆煲成湯，可補肺腎，適合秋冬季節飲用。

A soup made of lotus root, dried octopus and red bean can help nourish the Lungs and Kidneys, best to consume in autumn and winter.

材料

蓮藕 .... 1 斤
章魚 .... 2 兩
西施骨 .... 12 兩
紅豆 .... 2 兩
陳皮 .... 2/3 個

做法

❶ 陳皮用水浸 1 小時，刮淨
內瓤，洗淨。

❷ 紅豆洗淨，用水浸 1 小時，
隔去水分。

❸ 章魚洗淨，切段；蓮藕洗
淨污泥，切塊。

❹ 西施骨放入滾水內飛水 3
分鐘，撈起西施骨洗淨，
瀝乾水分。

❺ 蓮藕、紅豆、陳皮放入煲
中，注入水 12 碗煲滾，加
入西施骨、章魚煲滾，轉
中火煲 15 分鐘，再轉小火
煲 2 小時即成。

Ingredients

600 g lotus root
75 g dried octopus
450 g pork shoulder butt
75 g red beans
2/3 dried tangerine peel

Method

1. Soak dried tangerine peel in
water for 1 hour. Scrape off the
pith and rinse.

2. Rinse red beans, soak in water
for 1 hour, drain.

3. Rinse and cut dried octopus
into sections. Rinse lotus root
to clean off the mud. Cut into
thick pieces.

4. Scald pork shoulder butt in
boiling water for 3 minutes.
Dish up pork bone, rinse and
drain.

5. Add lotus root, red bean and
dried tangerine peel to pot.
Add 12 bowls of water. Bring
to boil. Add pork bone and
dried octopus. Bring to boil.
Adjust to medium heat and
cook for 15 minutes. Adjust
to low heat and simmer for 2
hours. Serve.

Luo Han Guo, Carrot and Dried White Cabbage Soup

羅漢果甘筍菜乾湯

● 化痰止咳、潤腸，有助排毒 ●

材料

羅漢果...........................半個
菜乾 ....1 兩
紅蘿蔔.........................12 兩
雪耳 ....半兩
南北杏.........................1 兩
陳皮 ....1/3 個

做法

❶ 陳皮用水浸軟，刮去內瓤。

❷ 雪耳用水浸泡1小時，剪去硬蒂，撕成細朵，飛水，過冷河，瀝乾水分。

❸ 菜乾用水浸泡2小時，洗淨砂粒，切短度；紅蘿蔔去皮，洗淨，切塊。

❹ 羅漢果及南北杏洗淨，隔乾水分。

❺ 煲滾清水10杯，放入全部材料，用大火煲滾，轉中火煲半小時，再轉小火煲1小時，下鹽調味即成。

小貼士

雪耳飛水及過冷河後，去掉
異味，而且口感爽脆好吃。

LUO HAN GUO, CARROT AND DRIED WHITE CABBAGE SOUP

expelling phlegm; stopping cough; nourishing the intestine; improving detoxifying

Ingredients

1/2 Luo Han Guo
38 g dried white cabbage
450 g carrot
19 g white fungus
38 g sweet and bitter almonds
1/3 dried tangerine peel

Method

1. Soak dried tangerine peel until soft. Scrape off the pith.
2. Soak white fungus for 1 hour. Scissor off the tough stems and tear into small pieces. Scald, rinse and drain.
3. Soak dried white cabbage for 2 hours. Rinse off the dirt and cut into short sections. Peel and rinse carrot. Cut into pieces.
4. Rinse Luo Han Guo and sweet and bitter almonds. Drain.
5. Bring 10 cups of water to boil. Put in all ingredients and bring to boil. Turn to medium heat and boil for half an hour. Turn to low heat and simmer for 1 hour. Season with salt. Serve.

TIPS

After scalding and rinsing, white fungus would become crispier and the unpleasant taste from white fungus are removed.

龍脷枇杷桔餅 杏仁瘦肉湯

● 化痰止咳、理氣，紓緩感冒後之咳嗽 ●

材料

乾龍脷葉 .... 1 兩
乾枇杷葉 .... 1 兩
桔餅 .... 2 個
南杏 .... 1 兩
北杏 .... 2 茶匙
陳皮 .... 1/3 個
瘦肉 .... 8 兩

做法

❶ 陳皮用水浸 1 小時，刮淨內瓤，
   洗淨。

❷ 瘦肉洗淨，切塊。

❸ 龍脷葉、枇杷葉、桔餅、南北杏
   全部洗淨，放在煲內，加入陳皮
   及水 9 碗煲滾，下瘦肉煲滾，轉
   小火再煲 1.5 小時，待暖飲用。

小貼士

乾龍脷葉清熱潤肺、止咳化痰，比鮮品較少寒涼；乾枇杷葉用於肺熱乾咳，常與乾龍脷葉一併使用；桔餅潤肺止咳，中藥店及乾果店有售。

LONG LI LEAF AND LOQUAT LEAF SOUP WITH ALMONDS AND LEAN PORK

Dissolving phlegm; stopping cough; regulating breathing; alleviating post-influenza cough

Ingredients

38 g dried Long Li leaf
38 g dried loquat leaf
2 preserved kumquats
38 g sweet almonds
2 tsp bitter almonds
1/3 dried tangerine peel
300 g lean pork

Method

1. Soak dried tangerine peel in water for 1 hour. Scrape off the pith and rinse.
2. Rinse and cut lean pork into pieces.
3. Rinse dried Long Li leaf, loquat leaf, preserved kumquat and almonds. Add to pot. Add dried tangerine peel and 9 bowls of water. Bring to boil. Add lean pork. Bring to boil. Adjust to low heat and simmer for 1.5 hours. Serve when soup becomes warm.

TIPS

Dried Long Li leaf can help clear Heat and sooth the Lungs, stop cough and dissolve phlegm. Its cooling nature is not as cold as that of the fresh leaf. Dried loquat leaf is often used together with dried Long Li leaf for relieving dry cough due to Heat from the Lungs. Preserved kumquat is good for nourishing the Lungs and relieving cough. It is sold in Chinese herbal shop or dried fruit store.

養好肺 • 強體質

Pork Soup with Coco-de-Mer, Carrot and Lily Bulb

海底椰甘筍百合瘦肉湯

● 滋潤肺部、止咳化痰 ●

moistening the Lungs; stopping cough; expelling phlegm

材料

乾海底椰 .... 1 兩
乾百合 .... 2 兩
甘筍 .... 1 條（約半斤）
瘦肉 .... 半斤
蜜棗 .... 2 粒
陳皮 .... 1/3 個

Ingredients

38 g dried coco-de-mer

75 g dried lily bulbs

1 carrot (about 300 g)

300 g lean pork

2 candied dates

1/3 dried tangerine peel

做法

❶ 陳皮用水浸軟，刮去瓤。

❷ 瘦肉洗淨，飛水，瀝乾水分。

❸ 甘筍去皮，洗淨，切塊。

❹ 海底椰、百合、蜜棗洗淨。

❺ 燒滾清水 12 杯，放入全部材料，大火煲 20 分鐘，轉小火煲 2 小時，下鹽調味。

Method

1. Soak dried tangerine peel until soft. Scrape off the pith.
2. Rinse, scald and drain lean pork.
3. Peel the carrot. Rinse and cut into wedges.
4. Rinse coco-de-mer, lily bulbs and candied dates.
5. Bring 12 cups of water to the boil. Put in all ingredients and boil over high heat for 20 minutes. Turn to low heat and simmer for 2 hours. Season with salt and serve.

小貼士 / TIPS

陳皮的內瓤帶苦澀味，煲湯時必須去掉。

We should scrape off the white pith from dried tangerine peel because it has a strong bitter taste.

Stewed Conch Soup with American Ginseng and Longan

花旗參桂圓燉螺頭湯

● 補肺氣、益氣血、健脾胃、滋陰生津 ●

nourishing the Qi in the Lungs; promoting Qi flow;
benefiting the Spleen and Stomach; nourishing the Yin;
stimulating body fluid secretion

小貼士 / TIPS

用牙刷徹底擦淨螺頭，去掉內臟並飛水。
You should clean conch with a toothbrush thoroughly. Discard the gut and scald well before cooking.

材料

豬腱 ....半斤
急凍螺頭 .... 10 兩
金華火腿 .... 1 兩
桂圓肉....................... 3/4 兩
杞子 ....1 湯匙
水 .... 5 杯

調味料

鹽.................................適量

做法

❶ 豬腱飛水後洗淨。
❷ 用牙刷擦乾淨急凍螺頭，
　棄去內臟，飛水，洗淨。
❸ 桂圓肉和杞子洗淨。
❹ 煲滾所有材料，轉放入大
　燉盅內，隔水燉 2 小時可
　品嘗，可下適量鹽調味。

Ingredients

300 g pork shin
375 g frozen conch
38 g Jinhua ham
30 g dried longan
1 tbsp Qi Zi
5 cups water

Seasoning

salt

Method

1. Scald and rinse pork shin.
2. Clean conch with a toothbrush and discard the gut. Scald and rinse.
3. Rinse dried longan and Qi Zi.
4. Bring all ingredients to boil and transfer in a big stewing pot. Stew for 2 hours. Season with salt. Serve.

Watercress and Duck Gizzard Soup with Chestnut and Pork Bone

西洋菜鴨腎栗子西施骨湯

清熱潤肺，紓緩口乾、喉嚨乾涸及便秘等

Clearing Heat; moistening the Lungs;
relieving problems such as thirst, dry throat and constipation

小貼士 / TIPS

西洋菜是冬天的蔬菜，止咳化痰、清燥潤肺。
Watercress is a winter vegetable that can help stop cough, dissolve phlegm, relieve dryness and moisten the Lungs.

材料

西洋菜............................ 1 斤
乾鴨腎............................ 2 個
栗子肉............................ 4 兩
西施骨............................ 12 兩
薑.................................... 3 片

做法

❶ 西洋菜用水浸半小時，再
　洗 3 次至西洋菜乾淨，瀝
　乾水分。

❷ 乾鴨腎洗淨，剕十字花；
　栗子肉洗淨。

❸ 西施骨放入滾水內飛水 3
　分鐘，取出，洗淨，瀝乾
　水分。

❹ 煲滾清水 12 碗，放入西施
　骨、乾鴨腎、西洋菜、薑
　片煲滾，轉中火煲 15 分
　鐘，轉小火再煲 1 小時，
　加入栗子肉煲 45 分鐘，一
　併食用湯料。

Ingredients

600 g watercress
2 dried duck gizzard
150 g shelled chestnuts
450 g pork shoulder butt
3 slices ginger

Method

1. Soak watercress in water for
   1/2 hour. Rinse three times to
   thoroughly clean watercress,
   drain.

2. Rinse dried duck gizzard,
   score shallow cross cuts on
   the surface. Rinse chestnuts.

3. Scald pork shoulder butt in
   boiling water for 3 minutes,
   rinse and drain.

4. Bring 12 bowls of water to boil.
   Add pork bone, dried duck
   gizzard, watercress and ginger.
   Bring to boil. Adjust to medium
   heat and cook for 15 minutes.
   Adjust to low heat and simmer
   for 1 hour. Add chestnuts and
   simmer for 45 minutes. Serve
   soup and the ingredients
   together.

Yacon Soup with Yellow Fungus and Black-eyed Bean

黃耳眉豆天山雪蓮素湯

● 潤肺、健脾胃 ●

nourishing the Lungs; strengthening the Spleen and Stomach

材料

黃耳 ....半兩
眉豆 .... 2 兩
花生 .... 2 兩
天山雪蓮 .... 1 斤
蜜棗 .... 2 粒
陳皮 ....半個

做法

❶ 黃耳用清水浸一晚，削去硬蒂，切小塊，飛水，過冷河，瀝乾水分。

❷ 陳皮用水浸軟，刮淨瓤。

❸ 眉豆及花生洗淨，用水浸 1 小時，瀝乾水分。

❹ 天山雪蓮洗淨污泥，去外皮，切大塊。

❺ 煲內注入清水 12 杯，下陳皮、眉豆、花生及蜜棗煲滾，加入黃耳及天山雪蓮煲滾，轉小火煲 1.5 小時，下鹽調味即可。

Ingredients

19 g yellow fungus
75 g black-eyed beans
75 g peanuts
600 g yacon
2 candied dates
1/2 dried tangerine peel

Method

1. Soak yellow fungus overnight. Cut off the hard stems and cut into small pieces. Scald, rinse and drain.
2. Soak dried tangerine peel until soft. Scrape off the pith.
3. Rinse black-eyed beans and peanuts. Soak for 1 hour. Drain.
4. Rinse off the mud from yacon. Peel and cut into large pieces.
5. Pour 12 cups of water in a pot. Put in dried tangerine peel, black-eyed beans, peanuts and candied dates. Bring to boil. Put in yellow fungus and yacon and bring to boil. Turn to low heat and boil for 1.5 hours. Season with salt. Serve.

小貼士 / TIPS

如沒有天山雪蓮及黃耳，可以
沙葛及雪耳代替，功效相同。
Replace yacon and yellow
fungus with yam beans and
white fungus; they have the
same functions.

Partridge Soup with Walnut and Gingko

銀杏合桃鷓鴣湯

● 強身健肺、理氣定喘，適合痰多咳嗽人士 ●

Strengthening body and the Lungs; regulating breathing;
soothing asthma; good for people who suffer from much phlegm and cough

小貼士 / TIPS

銀杏有定喘止咳的功能，加上潤肺化痰的合桃及花生，三者合用對秋燥咳嗽有幫助。

Ginkgo has been known to help calm asthma and relieve cough. Combining with walnut and peanut that are good for moistening the Lungs and dissolving phlegm, it is good for alleviating cough symptom caused by autumn dryness.

材料

```
冰鮮鷓鴣 ....................... 1 隻
瘦肉 ............................ 8 兩
銀杏 ............................ 2 兩
合桃肉 ......................... 2 兩
花生 ............................ 2 兩
陳皮 ......................... 2/3 個
```

做法

❶ 陳皮用水浸 1 小時，刮淨內瓤，洗淨。

❷ 銀杏用適量滾水浸 10 分鐘，褪去外衣，去芯，洗淨。

❸ 瘦肉洗淨，切塊。合桃、花生一同洗淨。

❹ 鷓鴣洗淨，放入滾水內飛水 3 分鐘，盛起，瀝乾水分。

❺ 銀杏、合桃肉、花生、陳皮放入煲內，注入水 10 碗煲滾，下鷓鴣、瘦肉煲滾，轉中火煲 15 分鐘，再轉小火煲 2 小時即成。

Ingredients

1 chilled partridge

300 g lean pork

75 g ginkgoes

75 g shelled walnuts

75 g peanuts

2/3 dried tangerine peel

Method

1. Soak dried tangerine peel in water for 1 hour. Scrape off the pith and rinse.

2. Soak ginkgoes in boiled water for 10 minutes, peel off the skin, remove the core and rinse.

3. Rinse and cut lean pork into pieces. Rinse walnuts and peanuts.

4. Rinse partridge, scald with boiling water for 3 minutes. Drain.

5. Add ginkgoes, walnuts, peanuts and dried tangerine peel to pot. Add 10 bowls of water. Bring to boil. Add partridge and lean pork. Bring to boil. Adjust to medium heat and cook for 15 minutes. Adjust to low heat and simmer for 2 hours. Serve.

養好肺 • 強體質

Pork Soup with White Fungus, Dried Pear and Lily Bulb

雪耳梨乾百合瘦肉湯

● 止咳化痰、清熱潤燥 ●

stopping cough; expelling phlegm; clearing Heat; moistening the body

材料

瘦肉 ....半斤
雪耳 ....半兩
雪梨乾.......................... 1 兩
乾百合........................... 1 兩
南北杏........................... 1 兩
蜜棗 .... 2 粒
陳皮 .... 1/3 個

做法

❶ 陳皮用水浸軟，刮去瓤。
❷ 雪耳用水浸軟，摘去硬蒂，洗淨，飛水，瀝乾水分。
❸ 瘦肉洗淨，飛水，瀝乾水分。
❹ 雪梨乾、百合、南北杏、蜜棗洗淨。
❺ 燒滾清水 12 杯，放入全部材料，用大火煲 20 分鐘，轉小火煲 2 小時，下鹽調味即成。

Ingredients

300 g lean pork
19 g white fungus
38 g dried pear
38 g dried lily bulbs
38 g sweet and bitter almonds
2 candied dates
1/3 dried tangerine peel

Method

1. Soak dried tangerine peel until soft. Scrape off the pith.
2. Soak white fungus until soft. Cut off the stems and rinse. Scald and drain well.
3. Rinse, scald and drain lean pork.
4. Rinse dried pear, dried lily bulbs, sweet and bitter almonds and candied dates.
5. Bring 12 cups of water to the boil. Put in all ingredients and boil over high heat for 20 minutes. Turn to low heat and simmer for 2 hours. Season with salt and serve.

小貼士 / TIPS

梨乾是鴨嘴梨切片曬乾而成，有清熱潤燥之功效。
Dried pear is made from sliced dried Ya-li pear. It expels Heat and moistens the body.

Flathead Fish Soup with Chinese Yam, Lotus Seeds and Lily Bulb

鮮淮山蓮子百合牛鰍魚湯

● 健脾補肺，助消化，適合全家飲用 ●

strengthening the Spleen; tonifing the Lungs;
improving the digestive system. This soup is suitable for the family.

材料

鮮淮山............................半斤
牛鰍魚......................... 12 兩
瘦肉 .... 4 兩
蓮子及乾百合 ....各 1 兩
薑 .... 3 片

做法

❶ 蓮子及百合洗淨，用水浸
1 小時，瀝乾水分。

❷ 鮮淮山削去外皮，洗淨，
切段；瘦肉洗淨，切厚片。

❸ 牛鰍魚劏好，洗淨，抹乾
水分，與薑片放入鑊內，
煎至兩面微黃色，隔油。

❹ 煲滾清水 10 杯，放入所有
材料用大火煲 15 分鐘，轉
中小火煲 1 小時，下鹽調
味即成。

Ingredients

300 g Chinese yam
450 g flathead fish
150 g lean pork
38 g lotus seeds
38 g dried lily bulb
3 slices ginger

Method

1. Rinse lotus seeds and dried lily
   bulb. Soak for 1 hour and drain.
2. Scrape off the skin from Chinese
   yam. Rinse and cut into sections.
   Rinse lean pork and cut into thick
   slices.
3. Cut open and gut flathead fish.
   Rinse and wipe dry. Fry in a wok
   with ginger until both sides become
   light brown. Remove and drain.
4. Bring 10 cups of water to boil.
   Put in all ingredients and boil over
   high heat for 15 minutes. Turn to
   medium-low heat and boil for 1
   hour. Season with salt. Serve.

小貼士 / TIPS

淮山性質平和，健脾補肺，
對體虛人士有幫助。
Huai Shan and Chinese yam
are neutral in nature, which
can strengthen the Spleen;
tonify the Lungs and benefit
people with weak body.

Coco-de-Mer and Tai Zi Shen Soup
with Carrot and Corn

紅蘿蔔粟米海底椰
太子參瘦肉湯

● 清熱除燥、潤肺止咳、補益脾胃，
提高免疫功能 ●

材料

紅蘿蔔.........................12 兩
粟米 ....2 條
海底椰.........................1 兩
太子參.........................半兩
瘦肉 ....8 兩
薑.................................2 片

做法

❶ 瘦肉洗淨，切塊。

❷ 紅蘿蔔去皮，洗淨，切塊；粟米
洗淨，切段。

❸ 海底椰、太子參洗淨，瀝乾水分。

❹ 紅蘿蔔、粟米、海底椰、太子參
及薑片放入煲內，加水 10 碗煮
滾，下瘦肉煲滾，轉小火煲 2 小
時即可。

小貼士

海底椰清熱除燥、潤肺止咳；
太子參補益脾胃，增強抗病
能力，緩解因氣候轉變的鼻
敏感徵狀。

COCO-DE-MER AND TAI ZI SHEN SOUP WITH CARROT AND CORN

clearing Heat and dryness; moistening the Lungs; stopping cough; nourishing the Spleen and Stomach; improving immunity

Ingredients

450 g carrot
2 ears sweet corn
38 g coco-de-mer
19 g Tai Zi Shen
300 g lean pork
2 slices ginger

Method

1. Rinse and cut lean pork into pieces.
2. Peel, rinse and cut carrot into pieces. Rinse and cut sweet corn into sections.
3. Rinse coco-de-mer and Tai Zi Shen, drain.
4. Add carrot, sweet corn, coco-de-mer, Tai Zi Shen and ginger to pot. Add 10 bowls of water. Bring to boil. Add lean pork. Bring to boil. Adjust to low heat and simmer for 2 hours. Serve.

TIPS

Coco-de-mer can help clear Heat and dryness, moisten the Lungs and stop cough. Tai Zi Shen has a warm but not heaty nature. It can help nourish the Spleen and Stomach, improve body's immunity and relieve allergic rhinitis caused by weather change.

Chuan Bei and Loquat Leaf Soup with Lily Bulb

川貝枇杷葉百合瘦肉湯

● 清熱潤肺、化痰止咳 ●

材料

川貝 .... 2/3 兩
乾百合 .... 2 兩
乾枇杷葉 .... 1.5 兩
陳皮 .... 2/3 個
瘦肉 ....半斤

做法

❶ 陳皮用水浸 1 小時，刮淨內瓤，洗淨。

❷ 瘦肉洗淨，切片。

❸ 川貝洗淨，拍裂。百合、乾枇杷葉洗淨。

❹ 川貝、百合、乾枇杷葉、陳皮放入煲內，加入水 10 碗煲滾，下瘦肉煲滾，轉小火煲 1 小時，待暖飲用。

小貼士

枇杷葉有清熱潤肺、化痰止咳的療效，搭配有利肺部及氣管的川貝、百合煲飲，增強呼吸系統。

CHUAN BEI AND LOQUAT LEAF SOUP WITH LILY BULB

clearing Heat; moistening the Lungs; dissolving phlegm; stopping cough

Ingredients

25 g Chuan Bei
75 g dried lily bulb
57 g dried loquat leaf
2/3 dried tangerine peel
300 g lean pork

Method

1. Soak dried tangerine peel in water for 1 hour. Scrape off the pith and rinse.
2. Rinse and slice lean pork.
3. Rinse and smash Chuan Bei. Rinse lily bulb and dried loquat leaf.
4. Put Chuan Bei, lily bulb, dried loquat leaf and dried tangerine peel in pot. Add 10 bowls of water. Bring to boil. Add lean pork. Bring to boil. Adjust to low heat and simmer for 1 hour. Serve warm.

TIPS

Loquat leaf can help clear Heat, moisten the Lungs, dissolve phlegm and stop cough. Combining with Chuan Bei and lily bulb that are good for strengthening the Lungs and respiratory tract, this soup can help strengthen the respiratory health.

養好肺 • 強體質

Chestnut, Job's Tears and Lily Bulb Soup

栗子薏米百合瘦肉湯

● 健脾益肺、養陰生津 ●

strengthening the Spleen; benefiting the Lungs;
nourishing the Yin; stimulating saliva secretion

材料

栗子 .... 6 兩（去殼）
瘦肉 .... 半斤
薏米 .... 1 兩
乾百合 .... 1 兩
南杏 .... 1 兩
陳皮 .... 1/3 個

Ingredients

225 g chestnuts (shelled)
300 g lean pork
38 g Job's tears
38 g dried lily bulbs
38 g sweet almonds
1/3 dried tangerine peel

做法

❶ 陳皮用水浸軟，刮去瓤。

❷ 栗子放入滾水內燙一會，取出去皮，洗淨。

❸ 瘦肉洗淨，飛水，瀝乾水分。

❹ 薏米、百合、南杏洗淨。

❺ 燒滾清水 12 杯，放入全部材料，大火煲 20 分鐘，轉小火煲 2 小時，下鹽調味。

Method

1. Soak dried tangerine peel until soft. Scrape off the pith.
2. Soak the chestnuts in boiling water for a while. Peel and rinse well.
3. Rinse, scald and drain lean pork.
4. Rinse Job's tears, lily bulbs and sweet almonds.
5. Bring 12 cups of water to the boil. Put in all ingredients and boil over high heat for 20 minutes. Turn to low heat and simmer for 2 hours. Season with salt and serve.

小貼士 / TIPS

用滾水輕燙栗子，再用乾毛巾包着，以雙手推擦數次，可輕易去掉栗子皮。

You can put chestnuts into boiling water for a while, wrap them with dried towel and rub around it. It can remove the chestnut skin easier.

養好肺・強體質

Fish Head Soup with Chinese Yam and Fox Nut

鮮淮山芡實魚頭湯

● 潤肺、健脾胃、治風虛頭痛 ●

moistening the Lungs; strengthening the Spleen and Stomach;
improving Wind-asthenia headache

小貼士 / TIPS

大魚頭用薑片略煎，去掉魚腥味，加上甘筍等材料同煲，湯水帶清甜味。

After frying the fish head with ginger, the unpleasant smell is already removed. Boiling with ingredients like carrot, the soup would taste good and sweet.

材料

大魚頭........................... 1 個
鮮淮山..........................半斤
芡實 .... 1 兩
甘筍 ....半斤
瘦肉 .... 4 兩
薑................................ 2 片

做法

1. 芡實洗淨，用水浸 1 小時，
   瀝乾水分。

2. 鮮淮山洗淨外皮，用小刀
   削去皮，洗淨，切塊，用
   水浸過面。

3. 甘筍去皮，洗淨，切塊；
   瘦肉洗淨，切片。

4. 大魚頭開邊，去鰓，洗淨，
   抹乾水分。

5. 燒熱鑊下少許油，放入薑
   1 片及大魚頭煎至魚頭微
   黃，隔油。

6. 煲內注入清水 8 杯煮滾，
   下所有材料用大火煲滾，
   轉中火再煲 40 分鐘，最後
   下鹽調味即成。

Ingredients

1 bighead carp head

300 g Chinese yam

38 g fox nuts

300 g carrot

150 g lean pork

2 slices ginger

Method

1. Rinse fox nuts. Soak for 1 hour and drain.

2. Rinse Chinese yam and scrape off the peel with a small knife. Rinse, cut into pieces and soak under water.

3. Peel carrot, rinse and cut into pieces; rinse and slice lean pork.

4. Cut fishheads into halves. Remove the gills, rinse and wipe dry.

5. Heat a wok with oil, fry fishheads with 1 slice of ginger until they turn light brown. Drain the oil.

6. Bring 8 cups of water to boil. Put in all ingredients and bring to boil over high heat. Turn to medium heat and boil for 40 minutes. Season with salt. Serve.

Asian Moon Scallop Soup with Water Chestnut and Carrot

沉魚落雁湯

● 補肝益腎、化痰潤肺 ●

strengthening the Liver and Kidneys;
expelling phlegm; nourishing the Lungs

材料

日月魚.......57 克（約 1.5 兩）
豬骨 .... 10 兩
馬蹄 ....5-6 粒
紅蘿蔔........................... 1 條
雪耳 .... 1/4 兩
水 .... 8 杯

調味料

鹽..................................適量

做法

❶ 豬骨洗淨，飛水。

❷ 日月魚洗淨後略浸。雪耳用水浸軟、去蒂，分拆小朵。

❸ 馬蹄去皮、洗淨。

❹ 紅蘿蔔去皮、洗淨，切件。

❺ 用大火煲滾所有材料，轉中慢火煲 2 小時，下適量鹽調味即成。

Ingredients

57 g Asian moon scallops
375 g pork bones
5-6 water chestnuts
1 carrot
10 g white fungus
8 cups water

Seasoning

salt

Method

1. Rinse and scald pork bones.
2. Rinse Asian moon scallops and soak briefly. Soak white fungus until soft. Tear off hard stems and tear into small pieces.
3. Peel and rinse water chestnuts.
4. Peel carrot, rinse and cut into pieces.
5. Bring all ingredients to boil over high heat. Turn to medium-low heat and simmer for 2 hours. Season with salt. Serve.

小貼士 / TIPS

日月魚用水略浸，煲湯後容易
散發香味。

Soaking Asian moon scallops
in water briefly, it is easier to
release the fragrant smell after
boiling.

Hairy Gourd, Lotus Seed and Lily Bulb Soup with Dried Oyster and Lean Pork

節瓜蓮子百合蠔豉瘦肉湯

● 開胃生津、清熱止渴、潤肺止咳 ●

improving appetite and saliva secretion; clearing Heat;
quenching thirst; nourishing the Lungs; stopping cough

小貼士 / TIPS

蓮子及百合是很好的養生食材,清心安神,促進新陳代謝;百合潤肺止咳,對身體有很大裨益。

Lotus seed and lily bulb are ingredients with great health benefits. Lotus seed can help calm nerves and emotion, improve metabolism. Lily bulb is good for nourishing the Lungs, stopping cough and great for improving health.

材料

節瓜 .... 1 斤
蠔豉 .... 2 兩
蓮子、乾百合 .... 各 1 兩
蜜棗 .... 3 粒
瘦肉 .... 8 兩
薑 .... 3 片

做法

❶ 蓮子、百合洗淨，用水浸 1 小時，隔去水分。

❷ 蠔豉洗淨，加入滾水浸 15 分鐘，隔去水分。

❸ 瘦肉洗淨，切塊；節瓜刮去外皮，洗淨，切段。

❹ 節瓜、蓮子、百合、蜜棗、薑片放入煲內，注入水 10 碗煲滾，加入蠔豉、瘦肉煲滾，轉中火煲 15 分鐘，再轉小火煲 1.5 小時，可連湯料一併食用。

Ingredients

600 g hairy gourd
75 g dried oysters
38 g lotus seed
38 g dried lily bulbs
3 candied dates
300 g lean pork
3 slices ginger

Method

1. Rinse lotus seed and lily bulb. Soak them in water for 1 hour, drain.

2. Rinse dried oysters. Soak in boiling water for 15 minutes, drain.

3. Rinse and cut lean pork into pieces. Peel hairy gourd, rinse and cut into sections.

4. Put hairy gourd, lotus seeds, lily bulb, candied dates and ginger into pot. Add 10 bowls of water. Bring to boil. Add dried oysters and lean pork. Bring to boil. Adjust to medium heat and cook for 15 minutes. Adjust to low heat and simmer for 1.5 hours. Serve the soup and the ingredients.

Partridge Soup with Long Li Leaves and Fig

龍脷葉花果鷓鴣湯

● 清熱潤肺、化痰止咳 ●

clearing Heat; nourishing the Lungs; expelling phlegm; stopping cough

材料

乾龍脷葉 .... 2 兩
無花果........................... 4 個
川貝 ....半兩
鷓鴣 .... 1 隻
瘦肉 .... 6 兩
陳皮 ....半個

做法

❶ 陳皮用水浸軟,刮淨內瓤。

❷ 鷓鴣洗淨,飛水,過冷河,
   瀝乾水分。

❸ 瘦肉洗淨,切厚片。

❹ 龍脷葉洗淨,瀝乾水分;
   無花果及川貝洗淨。

❺ 煲內注入清水 12 杯,加入
   陳皮煲滾,下其餘材料煲
   15 分鐘,轉小火煲 1.5 小
   時,下鹽調味享用。

Ingredients

75 g dried Long Li leaves

4 dried figs

19 g Chuan Bei

1 partridge

225 g lean pork

1/2 dried tangerine peel

Method

1. Soak dried tangerine peel until
   soft. Scrape off the pith.
2. Rinse partridge. Scald, rinse
   again and drain.
3. Rinse lean pork and cut into
   thick slices.
4. Rinse Long Li leaves and drain;
   rinse dried figs and Chuan Bei.
5. Pour 12 cups of water in a
   pot. Put in dried tangerine
   peel. Bring to boil. Put in other
   ingredients and boil for 15
   minutes. Turn to low heat and
   simmer for 1.5 hours. Season
   with salt. Serve.

小貼士 / TIPS

鷓鴣對脾虛、久咳很有幫助。現時市面只有急凍鷓鴣出售,或可用鵪鶉、瘦肉代替。

Partridges are very useful for Spleen deficiency and continuous cough. Only frozen partridges are available now. You can use quail or lean pork to replace.

White Fungus and Lily Bulb Soup with Dried Fig and Pork

雪耳百合無花果瘦肉湯

● 滋陰潤肺、有助便秘 ●

nourishing the Yin; moistening the Lungs; relieving constipation

小貼士 / TIPS

乾百合要選完整及體大的；雪耳可揀黃色、無經漂白的最佳。

Pick dried lily bulb that is intact and big in size. The best white fungus is the type that is yellow in colour, which has not gone through bleaching.

材料

雪耳 .... 1 球
乾百合 .... 2 兩
無花果 .... 4 個
陳皮 .... 1/3 個
瘦肉 .... 8 兩

做法

❶ 陳皮用水浸 1 小時，刮淨
內瓤，洗淨。

❷ 雪耳用水浸 1 小時至軟身，
剪去硬蒂，洗淨，飛水，
過冷河，瀝乾水分。

❸ 瘦肉洗淨，切厚片；百合、
無花果洗淨。

❹ 百合、無花果、陳皮放在
煲內，注入水 8 碗煲滾，
加入瘦肉煲滾，轉小火煲
45 分鐘，最後加入雪耳煲
半小時即成。

Ingredients

1 head white fungus
75 g dried lily bulbs
4 dried figs
1/3 dried tangerine peel
300 g lean pork

Method

1. Soak dried tangerine peel in water for 1 hour. Scrape off the pith and rinse.

2. Soak white fungus in water for 1 hour until tender, trim off the hard stem and rinse. Briefly scald white fungus with boiling water, rinse in cold water, drain.

3. Rinse and cut lean pork into thick slices. Rinse lily bulb and dried figs.

4. Add lily bulb, dried figs and dried tangerine peel to pot. Add 8 bowls of water. Bring to boil. Add lean pork. Bring to boil. Adjust to low heat and cook for 45 minutes. Add white fungus and simmer for 1/2 hour. Serve.

Tai Zi Shen Soup with Conch and Monkey Head Mushroom

太子參猴頭菇 豬腱螺片湯

● 益氣養陰、生津、補脾胃 ●

strengthening Qi; replenishing the Yin; stimulating saliva secretion; nourishing the Spleen and Stomach

小貼士 / TIPS

乾螺片滋陰補腎、清熱明目，而且不寒不燥，適合任何體質人士飲用。

Dried conch slice can tonify the Yin; strengthen the Kidneys; expel Heat and improve eyesight. It is neutral in nature and suitable for everyone.

材料

太子參........................ 1/3 兩
猴頭菇.......................... 1 兩
乾螺片.......................... 3 兩
豬腱 .... 12 兩
甘筍 ....半斤
薑................................ 2 片

做法

❶ 猴頭菇用水浸透（約 1 小時），剪去硬蒂，撕成小塊，洗淨，飛水，過冷河，擠乾水分。

❷ 豬腱洗淨，切大塊，飛水，過冷河，洗淨。

❸ 甘筍去皮，洗淨、切塊。

❹ 太子參及乾螺片同洗淨。

❺ 煲內注入清水 12 杯煲滾，放入全部材料用大火煲滾，續煲 15 分鐘，轉小火煲 1.5 小時，下少許鹽調味即成。

Ingredients

12 g Tai Zi Shen
38 g monkey head mushrooms
113 g dried conch slices
450 g pork shin
300 g carrot
2 slices ginger

Method

1. Soak monkey head mushrooms thoroughly (about 1 hour). Cut off hard stems and tear into small pieces. Rinse, scald, rinse again and squeeze dry.

2. Rinse pork shin and cut into large pieces. Scald and rinse.

3. Peel carrot. Rinse and cut into pieces.

4. Rinse Tai Zi Shen and conch.

5. Bring 12 cups of water and bring to boil. Put in all ingredients and bring to boil over high heat. Boil for 15 minutes. Turn to low heat and simmer for 1.5 hours. Season with salt. Serve.

Sugar Cane and Water Chestnut Soup
with Carrot and Pork

竹蔗馬蹄甘筍
瘦肉湯

● 解燥潤肺，紓緩秋燥引起之舌乾口渴 ●

Relieving dryness; moistening the Lungs; relieving symptoms of dry
tongue and thirst caused by the dry autumn weather

材料

竹蔗 .... 2 段
馬蹄 .... 10 粒
甘筍 ....半斤
瘦肉 ....半斤
陳皮 .... 1/3 個

做法

❶ 陳皮用水浸 1 小時，刮淨內瓤，洗淨。

❷ 竹蔗擦淨外皮，用刀破開成幼條。

❸ 馬蹄洗淨污泥，削去外皮，洗淨。

❹ 甘筍去外皮，洗淨，切塊。

❺ 瘦肉洗淨，切塊。

❻ 竹蔗、馬蹄、甘筍、陳皮放入煲內，注入水 10 碗煲滾，加入瘦肉煲滾，轉中火煲 15 分鐘，再轉小火煲 1.5 小時即成。

Ingredients

2 sections sugar cane
10 water chestnuts
300 g carrot
300 g lean pork
1/3 dried tangerine peel

Method

1. Soak dried tangerine peel in water for 1 hour. Scrape off the pith and rinse.

2. Rub and clean the skin of sugar cane. Cut the sugar cane into strips.

3. Rinse water chestnuts to clean off the mud. Peel and rinse.

4. Peel the carrot and rinse. Cut into pieces.

5. Rinse and cut lean pork into pieces.

6. Add sugar cane, water chestnuts, carrot and dried tangerine peel to pot. Add 10 bowls of water. Bring to boil. Add lean pork. Bring to boil. Adjust to medium heat and cook for 15 minutes. Adjust to low heat and simmer for 1.5 hours. Serve.

Pork Knuckle Soup with Duck Gizzard and Watercress

西洋菜鴨腎 豬蹄湯

● 潤肺化痰、清熱咳 ●

材料

乾鴨腎.............................. 2 個
豬蹄 .... 12 兩
西洋菜.............................. 1 斤
乾百合.............................. 1 兩
南北杏.............................. 1 兩
蜜棗 .... 3 粒
陳皮 ....半個

做法

❶ 陳皮用水浸軟，刮淨內瓤。

❷ 乾鴨腎洗淨，用水浸 1 小時，瀝乾水分。

❸ 豬蹄肉洗淨，飛水，過冷河，瀝乾水分。

❹ 西洋菜洗淨，用淡鹽水浸半小時，用水沖淨。

❺ 百合、南北杏及蜜棗同洗淨。

❻ 煲內注入清水 14 杯煲滾，放入豬蹄、乾鴨腎、蜜棗及陳皮，用中火煲 40 分鐘，再下西洋菜、百合及南北杏，大火煲滾，轉小火煲 1 小時，下鹽調味即可。

小貼士

市售的鴨腎已很潔淨，只要浸水去掉鹹味即可使用；最後下鹽調味前，應先試味，鴨腎之鹹香味或已令湯水濃郁可口。

PORK KNUCKLE SOUP WITH DUCK GIZZARD AND WATERCRESS

nourishing the Lungs; expelling phlegm;
relieving Heat-type coughing

Ingredients

2 dried duck gizzards

450 g pork knuckle

600 g watercress

38 g dried lily bulbs

38 g bitter and sweet almonds

3 candied dates

1/2 dried tangerine peel

Method

1. Soak dried tangerine peel until soft. Scrape off the pith.
2. Rinse dried duck gizzards. Soak for 1 hour and drain.
3. Rinse pork knuckle. Scald, rinse again and drain.
4. Rinse watercress, soak in lightly salted water for 30 minutes. Rinse.
5. Rinse lily bulbs, almonds and candied dates.
6. Bring 14 cups of water to boil. Put in pork knuckle, dried duck gizzards, candied dates and dried tangerine peel. Boil over medium heat for 40 minutes. Put in watercress, lily bulbs and almonds. Bring to boil over high heat. Turn to low heat and simmer for 1 hour. Season with salt. Serve.

TIPS

Today, available duck gizzards are clean enough; you should just soak them under water to remove excessive salty taste. Because of the salty taste of duck gizzards, you should taste the soup before seasoning with salt.

家常菜・粥

Homemade dishes & congee

將潤肺食材加入日常餐膳，
時時刻刻滋潤身體，保持健康的體魄！

Stir Fried Shrimps with Walnut, Lily Bulb and Ginkgoe

合桃百合銀杏炒蝦仁

● 滋養肺臟、增強免疫力 ●

moistening the Lungs; boosting the immunity

小貼士 / TIPS

如買回來的鮮蝦未烹調，建議用水浸泡，放於雪櫃冷藏，以免蝦頭及蝦身變黑。

It is suggest to soak the shrimps completely in water and put them in a refrigerator, or the shrimp head or body will turn black.

材料

烘香合桃肉.................. 12 粒
鮮百合 .... 2 球
銀杏 .... 20 粒
鮮蝦 ....半斤
乾葱 ....2 個（切片）
薑................................. 3 片
葱段 ....少許

醃料

胡椒粉...........................少許
粟粉 ....1 茶匙

調味料

蠔油 ....半湯匙

做法

❶ 鮮蝦去殼，挑腸，洗淨，
用廚房紙吸乾水分，下醃
料拌勻，放雪櫃冷藏備用。

❷ 鮮百合切去頭尾兩端，撕
成瓣狀，洗淨，瀝乾水分。

❸ 銀杏放入滾水焓 5 分鐘，
盛起。

❹ 燒熱鑊下油 2 湯匙，下薑
片、乾葱拌香，下蝦仁炒
勻，加入銀杏及熱水 2 湯
匙炒片刻，放入鮮百合炒
勻，下調味料、合桃肉、
葱段拌勻，上碟食用。

Ingredients

12 toasted skinned walnuts
2 fresh lily bulbs
20 ginkgoes
300 g fresh shrimps
2 shallots (sliced)
3 slices ginger
spring onion sections

Marinade

pepper
1 tsp cornflour

Seasoning

1/2 tbsp oyster sauce

Method

1. Shell and devein shrimps.
   Rinse and wipe dry any water
   with kitchen paper. Mix well
   with marinade and keep in the
   refrigerator.
2. Cut off both ends from lily bulbs
   and tear into pieces. Rinse and
   drain.
3. Boil gingkoes for 5 minutes.
   Remove.
4. Heat wok and add 2 tbsp of
   oil. Stir fry ginger and shallots
   until fragrant. Add shrimps
   and stir fry. Add gingkoes and
   2 tbsp of hot water and stir
   fry. Add lily bulbs and stir well.
   Mix in seasoning, walnuts and
   spring onion. Serve.

Fish Head Casserole with Garlic and Chinese Yam

蒜肉淮山魚頭鍋

● 健脾益胃、益肺止咳、殺菌 ●

strengthening the Spleen and Stomach; enriching the Lungs;
stopping cough; killing bacteria

小貼士 / TIPS

鮮淮山可炒吃、煲粥或製成甜品，去皮後宜泡水以免氧化。
Chinese yam is used to make stir-fried dishes, congees and desserts.
To prevent oxidation after peeling, it should be soaked in water.

材料

大魚頭.........1 個（約 12 兩）
鮮淮山...........................6 兩
瘦肉 ....4 兩
炒香蒜肉 ....6 粒
薑 ....3 片

醃料

鹽.............................半茶匙

做法

❶ 大魚頭開邊，去掉魚鰓（請
魚販代勞），洗淨及抹乾
水分，抹上醃料待片刻。

❷ 鮮淮山削皮，洗淨，切片，
用清水浸片刻，盛起。

❸ 瘦肉洗淨，切薄片。

❹ 燒熱鍋下油 1 湯匙，下薑
片及大魚頭煎至微黃，注
入熱水 2 杯煮 10 分鐘，
加入瘦肉、淮山及蒜肉，
用慢火煮 10 分鐘，原鍋上
桌食用。

Ingredients

1 big fish head (about 450 g)
225 g Chinese yam
150 g lean pork
6 cloves toasted peeled garlic
3 slices ginger

Marinade

1/2 tsp salt

Method

1. Slit the side of big fish head to remove the gills (request for the help of fishseller). Rinse fish head. Pat dry. Brush on marinade. Let stand for a while.

2. Peel Chinese yam. Rinse and slice. Soak in water for while. Dish up.

3. Rinse and thinly slice lean pork.

4. Heat wok. Add 1 tbsp of oil. Add ginger and big fish head. Pan-fry until fish head turns golden. Add 2 cups of hot water. Cook for 10 minutes. Add lean pork, Chinese yam and garlic. Cook over low heat for 10 minutes. Serve in casserole.

養好肺 • 強體質

Fish and White Fungus Couscous Porridge

魚茸雪耳小米粥

● 潤肺、健脾固腎，滋潤身體 ●

nourishing the Lungs; strengthening the Spleen and Kidneys;
moistening the body

小貼士 / TIPS

青根魚、黃花魚、紅衫魚的魚肉可製成魚茸，魚肉滑嫩，味道鮮甜，
價錢相宜。

The meat of piceus, yellow croaker or golden threadfin bream can be
used as fish flakes. The meat is tender and tasty. Price is reasonable.

材料

青根魚............1 條（約 8 兩）
雪耳 .... 1/4 球
小米 .... 1/3 量杯 *
白米 .... 1/3 量杯 *
薑 .... 3 片
鹽 ....1/3 茶匙
* 專用量米杯

做法

❶ 白米洗淨，用水浸 1 小時，
  瀝乾水分。

❷ 雪耳用水浸 1 小時，剪去
  硬蒂，飛水，撕成小朵。

❸ 青根魚洗淨，放於蒸碟，
  鋪上薑片，隔水大火蒸 10
  分鐘，待冷，拆出魚肉。

❹ 煮滾清水 7 碗，放入白米
  及小米煮滾，轉小火煮約
  半小時，加入雪耳再煮 5
  分鐘，下鹽調味，最後拌
  入魚茸即可食用。

Ingredients

1 piceus (about 300 g)
1/4 head white fungus
1/3 measuring cup couscous *
1/3 measuring cup white rice *
3 slices ginger
1/3 tsp salt
* Specific rice measuring cup

Method

1. Rinse white rice. Soak in water
   for 1 hour. Drain.
2. Soak white fungus in water
   for 1 hour. Trim away the hard
   stem. Briefly scald with boiling
   water. Drain and tear it into
   small pieces.
3. Rinse fish. Place fish on
   steaming plate. Top with ginger
   slices. Steam over water with
   high heat for 10 minutes. Let
   cool. Flake fishmeat and set
   aside.
4. Bring 7 bowls of water to boil.
   Add white rice and couscous.
   Bring to boil. Adjust to low
   heat and simmer for 1/2 hour.
   Add white fungus and cook for
   another 5 minutes. Add salt
   as seasoning. Lastly put in fish
   flakes. Stir well. Serve.

養好肺 • 強體質

113

Stir Fried Chicken, Chinese Yam and Bell Peppers

鮮淮山甜椒
炒雞片

● 補肺益腎、補氣 ●

tonifying the Lungs and the Kidneys; nourishing Qi

材料

鮮淮山............1 段（約 6 兩）
青、紅、黃甜椒 .... 各半個
雞柳 .... 3 兩
蒜茸 ....2 茶匙

醃料

生抽 ....1 茶匙
胡椒粉...........................少許
粟粉 ....半茶匙
油 .... 1 湯匙（後下）

調味料

蠔油 .... 半湯匙
鹽.............................. 半茶匙
水 ....3 湯匙

做法

❶ 雞柳洗淨，抹乾水分，切
片，下醃料拌勻待半小時。

❷ 鮮淮山洗淨，去皮，切薄
片，用水浸過面待半小時，
隔水。

❸ 甜椒去蒂、去籽，洗淨，
切塊。

❹ 燒熱鑊下油 2 湯匙，下甜
椒炒片刻，盛起。

❺ 原鑊下蒜茸及雞柳炒片
刻，加入鮮淮山、甜椒及
調味料炒片刻，至雞柳熟
透即成。

Ingredients

1 section Chinese yam (about 225 g)
1/2 green bell pepper
1/2 red bell pepper
1/2 yellow bell pepper
113 g chicken fillet
2 tsp grated garlic

Marinade

1 tsp light soy sauce
pepper
1/2 tsp cornflour
1 tbsp oil (added last)

Seasoning

1/2 tbsp oyster sauce
1/2 tsp salt
3 tbsp water

Method

1. Rinse chicken fillet and wipe dry. Slice and mix with marinade. Marinate for half an hour.
2. Rinse yam. Remove the skin and thinly slice. Soak in water (covered with the water) for half an hour. Drain.
3. Remove the stalks and seeds from bell peppers. Rinse and cut into pieces.
4. Heat wok and add 2 tbsp of oil. Stir fry bell peppers and set aside.
5. Stir fry garlic and chicken fillet in the same wok. Add yam, bell peppers and seasoning. Stir fry until chicken are fully done. Serve.

Millet Congee with Qi Zi Meatballs and Yam

杞子肉丸鮮淮山小米粥

● 補肺、益脾胃 ●

nourishing the Lungs; tonifying the Spleen and Stomach

小貼士 / TIPS

切勿大火時加入肉丸，要小火慢煮，並且慢慢攪動，以免黏着鍋底。

Do not use high heat to cook the meatballs; cook them slowly over low heat, stir lightly to prevent them from sticking to the pot.

材料

小米 .... 半量杯 *
白米 .... 半量杯 *
新鮮淮山 .... 4 兩
免治豬肉 .... 6 兩
杞子 .... 2 湯匙
沖菜 .... 半片
薑 .... 2 片
鹽 .... 半茶匙
*專用量米杯

調味料

生抽 .... 1 茶匙
胡椒粉 .... 少許
粟粉 .... 2 茶匙

做法

❶ 沖菜洗淨,切碎,與免治
豬肉、杞子、調味料拌至
起膠,用鐵匙弄成肉丸,
備用。

❷ 白米洗淨,用水浸 1 小時,
瀝乾水分。

❸ 淮山去皮,洗淨,切厚片,
用水浸泡。

❹ 煲滾清水 9 杯,加入白米、
小米、淮山煮滾,轉小火
煲半小時,待粥底綿滑,
加入杞子豬肉丸煮 10 分
鐘,最後下鹽調味即成。

Ingredients

1/2 measuring cup millet *
1/2 measuring cup white rice *
150 g Chinese yam
225 g minced pork
2 tbsp Qi Zi
1/2 slice preserved mustard green
2 slices ginger
1/2 tsp salt
* Specific rice measuring cup

Seasoning

1 tsp light soy sauce
pepper
2 tsp cornflour

Method

1. Rinse preserved mustard green, finely chop. Mix preserved mustard green, minced pork, Qi Zi and seasoning together. Stir until the mixture becomes sticky. Make meatballs with a tablespoon. Set aside.

2. Rinse white rice, soak for 1 hour and drain.

3. Skin Chinese yam, rinse and cut into thick slices. Soak in water.

4. Bring 9 cups of water in a pot to boil. Add white rice, millet, Chinese yam. Bring to boil. Turn to low heat and simmer for 30 minutes, until the congee is smooth. Add Qi Zi meatballs and cook for 10 minutes. Season with salt. Serve.

養好肺 • 強體質

Yam and Chicken Nabemono with Soybean Milk

豆漿鮮淮山雞肉鍋

● 潤肺、養陰生津，增強抵抗力 ●

nourishing the Lungs; tonifying the Yin; stimulating body fluid secretion; strengthening the immunity

小貼士 / TIPS

鮮淮山的黏液是精華所在，切勿洗掉。
Do not wash the fluid of Chinese yam as it is the nutrition essence.

材料

淡豆漿........................ 1.5 杯
雞肉 .... 6 兩
本菇 .... 1 包
鮮淮山......................... 3 兩
薑................................. 4 片
昆布 ....數塊
鰹魚片 ....2 湯匙
鹽............................適量

做法

❶ 鰹魚片用滾水半杯焗 10 分
鐘,隔出湯汁,備用。

❷ 本菇切去末端,洗淨;鮮
淮山去皮,洗淨,切片;
雞肉洗淨、切塊;昆布用
濕毛巾抹淨。

❸ 淡豆漿及鰹魚湯汁傾入鍋
內,加入昆布及薑片煲滾,
下雞肉及淮山煮 10 分鐘,
加入本菇煮片刻,最後下
適量鹽調味。

Ingredients

1.5 cups light soybean milk
225 g chicken meat
1 packet Hon-shimeji mushrooms
113 g Chinese yam
4 slices ginger
a couple piece kelp
2 tbsp dried bonito flakes
salt

Method

1. Soak the dried bonito flakes in 1/2 cup of boiling water covered with a lid for 10 minutes. Filter the soup. Set aside.

2. Cut away the root of the Hon-shimeji mushrooms. Rinse well. Peel, rinse and slice the Chinese yam. Rinse the chicken meat. Cut into pieces. Clean the kelp with a damp towel.

3. Pour the light soybean milk and bonito soup from step 1 into a saucepan. Add the kelp and ginger. Bring to the boil. Add the chicken meat and Chinese yam. Cook for 10 minutes. Put in the mushrooms and cook for a moment. Season with salt at last and serve.

Brown Rice with Porcini

牛肝菌糙米飯

● 抵抗流感、防感冒、增強免疫力 ●

fighting and preventing flu; strengthening the immunity

小貼士 / TIPS

先煮糙米，因為它需要較長時間才能煮腍；乾牛肝菌可在大型超市選購。

Brown rice must be cooked first because it takes longer time to get it done. Porcini can be bought at large supermarkets.

材料

糬米 ....1/4 量杯
白米 .... 半量杯
上湯 .... 半杯
乾牛肝菌 .... 3/4 兩
橄欖油......................1/2 湯匙

調味料

鹽.........................1/2 茶匙
黑椒碎.........................少許

做法

❶ 乾牛肝菌用水浸 10 分鐘盛
　起，略沖後再用水半杯浸
　約 20 分鐘，盛起，擠乾水
　分，水留起。

❷ 糬米、白米分別洗淨。

❸ 燒熱橄欖油，將牛肝菌略
　炒，盛起。

❹ 糬米先用水 1 杯煮約 20
　分鐘，然後加入白米和上
　湯半杯，繼續煮 10 分鐘，
　加入牛肝菌和調味料再煮
　片刻，慢火略焗約 20 分鐘
　便可以享用。

Ingredients

1/4 measuring cup brown rice
1/2 measuring cup white rice
1/2 cup stock
30 g dried porcini
1/2 tbsp olive oil

Seasoning

1/2 tsp salt
a pinch ground black pepper

Method

1. Sock porcini for 10 minutes.
   Remove and rinse for a while.
   Soak in 1/2 cup of water for
   20 more minutes. Remove and
   squeeze out the water. Reserve
   the liquid from soaking porcini.

2. Wash brown rice and white
   rice separately.

3. Heat olive oil. Stir fry porcini for
   a while. Remove.

4. Cook brown rice in 1 cup of
   water for 20 minutes. Add
   white rice and 1/2 cup of stock.
   Cook for 10 more minutes. Put
   in porcini and seasoning and
   cook for a while. Simmer over
   low heat for 20 minutes. Serve.

Pork Fillet with Chinese Yam and King Oyster Mushroom

鮮淮山杏鮑菇炒柳梅

● 健脾、益肺氣 ●

strengthening the Spleen; enriching Qi in the Lungs

小貼士 / TIPS

為了易於咀嚼，肉片、杏鮑菇及淮山等宜切成片狀。

To allow easy chewing, cut meat, king oyster mushroom and Chinese yam into slices.

材料

鮮淮山 .... 4 兩
杏鮑菇 .... 2 塊
柳梅肉 .... 4 兩
葱段 ....適量
薑 .... 3 片
乾葱 .... 1 粒（去衣、切碎）
紹酒 ....半湯匙

醃料

生抽 ....2 茶匙
粟粉 ....1 茶匙
水 ....2 湯匙

調味料

蠔油 .... 半湯匙

做法

❶ 柳梅肉洗淨，切片，下醃料拌勻待半小時。

❷ 鮮淮山洗淨污泥，削去外皮，切薄片，用清水浸半小時，瀝乾水分。

❸ 杏鮑菇洗淨，切片。

❹ 燒熱鑊下油 2 湯匙，下薑片及乾葱炒香，加入柳梅肉炒勻，灒酒炒勻，放入杏鮑菇、鮮淮山、熱水 3 湯匙，加蓋煮 5 分鐘，最後下調味料及葱段拌勻，上碟享用。

Ingredients

150 g Chinese yam
2 king oyster mushrooms
150 g pork fillet
spring onion sections
3 slices ginger
1 clove shallot (peeled and finely chopped)
1/2 tbsp Shaoxing wine

Marinade

2 tsp light soy sauce
1 tsp cornflour
2 tbsp water

Seasoning

1/2 tbsp oyster sauce

Method

1. Rinse and slice pork fillet. Combine with marinade ingredients and let stand for 1/2 hour.
2. Rinse Chinese yam to thoroughly clean off the mud. Peel the skin and thinly slice. Soak in water for 1/2 hour and drain.
3. Rinse and slice king oyster mushrooms.
4. Heat wok. Add 2 tbsp oil. Add ginger slices and shallot. Stir-fry until fragrant. Add pork fillet. Stir-fry until even. Sprinkle with wine and stir well. Add king oyster mushroom, yam and 3 tbsp hot water. Cover the lid and cook for 5 minutes. Add seasoning and spring onion sections. Stir until even. Serve.

Watercress Pork Dumpling

西洋菜豬肉餃子

● 清熱、解燥、潤肺 ●

材料

西洋菜	半斤
免治豬肉	半斤
圓形餃子皮	半斤

調味料

生抽	半湯匙
胡椒粉	少許
雞蛋	1 個
粟粉	1 湯匙

蘸汁

鎮江醋	1 湯匙
生抽	1 茶匙
麻油	2 茶匙

做法

❶ 西洋菜洗淨，放入滾水略灼，過冷河，擠乾水分，切粒，再次擠乾水分，備用。

❷ 免治豬肉、調味料、雞蛋攪拌均勻，加入西洋菜碎拌勻成餡料。

❸ 每塊餃子皮中間放入 1 湯匙餡料，於餃子皮內側塗抹少許水，按實即可。

❹ 煮滾清水半鍋，放入餃子用中火煮 8 分鐘，上碟，伴蘸汁食用。

小貼士

選幼嫩的西洋菜做成餃子；包餡料前可先冷藏，以免釋出水分。

WATERCRESS PORK DUMPLING

clearing Heat; relieving dryness; moistening the Lungs

Ingredients

300 g watercress
300 g minced pork
300 g round shaped dumpling wrapper

Seasoning

1/2 tbsp light soy sauce
pepper
1 egg
1 tbsp cornflour

Dipping sauce

1 tbsp Zhenjiang vinegar
1 tsp light soy sauce
2 tsp sesame oil

Method

1. Rinse and briefly blanch watercress with boiling water. Rinse with cold water. Squeeze dry and dice. Squeeze dry again.
2. Combine minced pork, seasoning ingredients and egg. Add diced watercress. Mix well as filling.
3. Place 1 tbsp of filling at the center of each dumpling wrapper. Wet the inside edge of dumpling wrapper with a little water. Pinch to seal.
4. Bring half pot of water to boil. Put in dumplings and cook over medium heat for 8 minutes. Transfer to plate. Serve with dipping sauce.

TIPS

The tender young watercress can be used for making dumplings. The filling should be chilled in a refrigerator before the wrapping to minimize the watery texture.

潤肺甜點

Nourishing sweets

杏汁、川貝、雪耳、楊桃，
製成多款滋潤甜點，甜心又潤身！

Papaya and White Fungus with Almond Milk

杏汁木瓜雪耳

● 止咳平喘、潤澤肌膚 ●

stopping cough; alleviating difficulty in breathing; nourishing complexion

小貼士 / TIPS

平和的杏仁是止咳平喘的常用食材，多用作煲湯及製成甜品，在乾燥的秋季飲用，有預防咳嗽及滋潤肌膚之效。

Almond kernel is commonly used for stopping cough and alleviate breathing difficulty. It is often used for making soup and dessert. Regular consumption during the dry autumn will help prevent cough and nourish complexion.

材料

熟木瓜肉 .... 半磅
南杏 .... 2 兩
北杏 ....1 湯匙
雪耳 .... 半球
冰糖 .... 2 兩
水 .... 6 杯
煲魚湯袋 .... 1 個

做法

❶ 雪耳用水浸 1 小時，摘去
蒂，撕成細朵，飛水，過
冷河，瀝乾水分。

❷ 南北杏洗淨，放入攪拌機，
加入水 2 杯磨成幼滑杏仁
漿，傾入煲魚湯袋內隔去
杏仁渣，過濾成杏仁漿。

❸ 木瓜肉切細塊；冰糖洗淨。

❹ 煮滾 4 杯水，放入冰糖、
雪耳煲滾，下木瓜煲滾，
傾入杏仁漿攪勻，用小火
慢煮至微滾，再煮 5 分鐘
即成。

Ingredients

225 g ripe papaya pulp

75 g sweet almonds

1 tbsp bitter almonds

1/2 head white fungus

75 g rock sugar

6 cups water

1 fish soup bag

Method

1. Soak white fungus in water for 1 hour. Remove the stem and tear white fungus into small pieces. Briefly blanch with boiling water. Rinse with cold water and drain.

2. Rinse sweet and bitter almonds. Pour into blender. Add 2 cups of water. Blend into almond milk. Pour into fish soup bag and strain. Set aside almond milk.

3. Cut papaya pulp into small pieces. Rinse rock sugar.

4. Bring 4 cups of water to boil. Add rock sugar and white fungus. Bring to boil. Add papaya. Bring to boil. Add almond milk and stir well. Adjust to low heat and bring to gentle boil. Cook for another 5 minutes. Serve.

養好肺・強體質

Ya-li Pear and Chuan Bei Soup

川貝煲雪梨

● 滋陰潤燥，對熱咳有改善作用 ●

nourishing the Yin; moistening dryness; alleviating Heat cough

Ingredients

30 g Chuan Bei

4 Ya-li pears

1 tbsp rock sugar (or 2 tbsp honey)

Method

1. Rinse and smash Chuan Bei.
2. Rinse pears, peel and cut into quarters. Discard stem and seeds.
3. Add 8 cups of water to a small pot. Add pears and Chuan Bei. Bring to boil. Adjust to low heat and simmer for 45 minutes. Add rock sugar and cook until sugar dissolves. Serve.

材料

川貝 .... 8 錢
鴨咀梨........................... 4 個
冰糖 ..1 湯匙（或蜜糖 2 湯匙）

做法

❶ 川貝洗淨，拍裂。

❷ 鴨咀梨洗淨外皮，切成 4 件，去蒂、去籽。

❸ 小煲內注入清水 8 杯，放入鴨咀梨、川貝煲滾，轉小火煲 45 分鐘，加入冰糖煮至溶化即可。

小貼士

川貝配搭雪梨有滋陰潤燥之效。此茶帶甜味，可減低川貝的微苦。

Combining with Ya-li pear and Chuan Bei, the soup can nourish the Yin and moisten dryness. The sweetness of this remedy can reduce the slight bitterness of Chuan Bei.

Creamy Peanut and Cashew Paste

花生腰果露

● 潤肺潤燥、滋補腰腎 ●

nourishing the Lungs and dryness; tonifying the Kidneys and waist

小貼士 / TIPS

淡奶是經過蒸餾的牛奶，水分比鮮奶少，適合烹調奶露等食品，而且奶味香濃！

Evaporated milk is distilled milk with water less than that in fresh milk. It is suitable for cooking milk paste giving an intense fragrance of milk.

材料

花生 .... 4 兩
腰果 .... 2 兩
淡奶 ....2 湯匙
冰糖 ....2 湯匙（舂碎）

粟粉獻

粟粉 ....1 湯匙
水 ....4 湯匙
*拌勻

做法

❶ 花生及腰果洗淨，放入白
　鑊用小火炒至表面呈金黃
　色，盛起；花生去皮，備
　用。

❷ 花生、腰果及水 2 杯，放
　入攪拌機內，磨成幼滑漿，
　用隔篩過濾。

❸ 煮滾清水 2 杯，下冰糖及
　花生腰果漿，邊下邊拌，
　煮至微滾片刻，加入粟粉
　獻拌勻煮滾，最後下淡奶
　攪拌即可。

Ingredients

150 g peanuts

75 g cashews

2 tbsp evaporated milk

2 tbsp rock sugar (crushed)

Thickening glaze

1 tbsp cornflour

4 tbsp water

* mixed well

Method

1. Rinse the peanuts and cashews. Stir-fry in a dry wok over low heat until the surface is golden. Remove. Skin the peanuts and set aside.

2. Put the peanuts, cashews and 2 cups of water into a blender. Grind into smooth batter. Sieve.

3. Bring 2 cups of water to the boil. Put in the rock sugar and peanut cashew batter. Keep stirring while putting. When it starts to boil, cook for a while. Stir in the thickening glaze. Bring to the boil. Finally fold in the evaporated milk. Serve hot.

養好肺 • 強體質

Steamed Brown Sugar Bowl Cakes with Red Date Shreds

紅棗茸黃糖缽仔糕

● 生津潤肺、補脾胃，調養體質 ●

stimulating body fluid secretion; nourishing the Lungs;
tonifying the Spleen and Stomach; strengthening the body

小貼士 / TIPS

粘米漿必須呈稀糊狀，蒸出來的缽仔糕才軟滑細緻；若粘米漿未呈
稀糊狀，可用小火拌煮至稀糊即可。

The rice batter must be thin so that the cake steamed is soft and
delicate. If the batter is not thin enough, cook over low heat and stir
until it turns thin.

材料

紅棗 .... 8 粒
粘米粉 .... 3.5 兩
澄麵 .... 2 兩
片糖 .... 6 兩
滾水 .... 3.5 杯

做法

❶ 紅棗洗淨，去核，切碎。

❷ 燒滾水 1 杯，放入片糖煮
滾，加入紅棗碎煲 5 分鐘，
待涼。

❸ 粘米粉及澄麵篩勻，傾入
紅棗糖水拌勻，再加滾水
2.5 杯（邊加入邊拌勻），
拌成稀糊漿。

❹ 將熱稀糊漿盛入小碗內，
隔水大火蒸 15 分鐘，待
暖，用竹籤挑出供食。

Ingredients

8 red dates
132 g rice flour
75 g Tang flour
225 g raw cane sugar slab
3.5 cups boiling water

Method

1. Rinse and core the red dates.
   Finely chop.

2. Bring 1 cup of water to the
   boil. Put in the cane sugar
   slab. Bring to the boil. Add the
   chopped red dates and cook
   for 5 minutes. Let it cool down.

3. Sieve the rice flour and Tang
   flour. Pour in the red date
   sweet soup from step 2. Mix
   well. Add 2.5 cups of boiling
   water (keep stirring while
   adding the water). Mix into
   thin batter.

4. Put the hot batter into small
   bowls. Steam over high heat
   for 15 minutes. Pick with
   bamboo skewers while warm.
   Serve.

Almond Tea with Egg White

蛋白杏仁茶

● 養顏、潤肺、化痰止咳 ●

improving skin texture; nourishing the Lungs;
expelling phlegm; stopping cough

材料

大南杏........................... 2 兩
北杏 ....1 湯匙
白米 ....1 湯匙
雞蛋白........................... 2 個
冰糖 ....1 湯匙

做法

❶ 大南杏、北杏及白米洗淨，
用水 1 杯浸 1 小時，隔去
水分。

❷ 將南杏、北杏及白米放入
攪拌機內，加入清水 1 杯
磨成幼漿，過濾，杏仁米
漿 1 杯留用。

❸ 煮滾清水 1 杯，傾入杏仁
米漿攪勻，煮至微滾，下
冰糖煮至溶化，待杏仁米
漿煮成稀糊狀，關火，傾
入蛋白待 1 分鐘，輕拌即
成。

Ingredients

75 g large sweet almonds
1 tbsp bitter almonds
1 tbsp rice
2 egg whites
1 tbsp rock sugar

Method

1. Rinse sweet almonds, bitter almonds and rice. Soak in 1 cup of water for 1 hour. Strain.
2. Put the sweet almonds, bitter almonds and rice into a blender. Add 1 cup of water. Grind into smooth batter. Filter and reserve 1 cup of the almond rice batter.
3. Bring 1 cup of water to the boil. Pour in the almond rice batter. Mix well. Cook until it slightly boils. Add the rock sugar and cook until it dissolves. When the batter turns into a thin paste. Turn off heat. Pour in the egg whites and wait for 1 minute. Gently mix to finish.

小貼士 / TIPS

緊記關火後才傾入蛋白，輕拌一會，
待蛋白慢慢凝固即成嫩滑蛋白。
Remember to turn off heat before
pouring in the egg white. Mix gently for
a while. When it slowly sets, the creamy
egg white is done.

養好肺 • 強體質

Double-steamed White Fungus with Sugarcane Juice

蔗汁燉雪耳

● 潤肺、益氣健脾 ●

nourishing the Lungs; benefiting Qi and the Spleen

材料

雪耳 .... 2 球
鮮榨蔗汁 .... 2 杯
冰糖 .... 1 粒

做法

❶ 雪耳用水浸 1 小時，剪去
硬蒂，摘細朵，飛水，過
冷河，瀝乾水分。

❷ 雪耳及冰糖置於燉盅內，
注入 1.5 杯滾水，隔水中
火燉 40 分鐘，加入蔗汁再
燉 20 分鐘即可。

Ingredients

2 heads white fungus
2 cups freshly pressed sugarcane
  juice
1 cube rock sugar

Method

1. Soak white fungus for 1 hour.
   Cut off hard stems and tear
   into small pieces. Scald, rinse
   and drain.
2. Put white fungus and rock
   sugar in a stewing pot. Pour
   in 1.5 cups of boiling water.
   Double-steam the pot for 40
   minutes. Pour in sugarcane
   juice and double-steam for 20
   minutes. Serve.

小貼士 / TIPS

熱飲的蔗汁有補益功效，潤燥止渴、益氣健脾，
適合肺熱乾咳者飲用。

Hot sugarcane juice has a great invigorating
function of nourishing, stopping thirst and
strengthening the Spleen. It is suitable for people
with Heat in the Lungs and dry cough.

Creamed White Fungus Sweet Soup with Huai Shan

淮山雪耳奶露

● 補腎益氣、強身健體 ●

benefiting the Kidneys; promoting Qi (vital energy);
strengthening the body

材料（份量：500 毫升）

淮山 ....5 兩 3 錢
雪耳 .... 1 錢
脫脂奶.......................... 2 杯
蜂蜜 ....1 湯匙

做法

❶ 淮山去皮、切薄片；雪耳浸軟後去蒂。

❷ 燒滾 1 杯水，下淮山、雪耳，煮約 20 分鐘，關火，焗至軟透。

❸ 將所有材料放入攪拌機內，攪至幼滑便可以飲用。

Ingredients (yields 500ml)

200 g Huai Shan

4 g white fungus

2 cups skimmed milk

1 tbsp honey

Method

1. Peel the Huai Shan and slice thinly. Soak the white fungus in water until soft. Cut off the yellow stems.

2. Boil 1 cup of water. Put in Huai Shan and white fungus. Boil for 20 minutes. Turn off the heat and cover the lid. Leave it in the pot until ingredients are tender.

3. Put all ingredients into a blender. Puree until smooth. Serve.

小貼士 / TIPS

雪耳必須浸軟，切去硬蒂才
煲煮。

You should soaked white
fungus in water until soft,
cutting off the hard stem
before cooking.

養好肺 • 強體質

Peach Resin Sweet Soup with Lotus Seed and Lily Bulb

桃膠蓮子百合糖水

● 潤肺、滋潤美顏、嫩膚補身 ●

nourishing the Lungs; improving complexion; softening the skin;
nourishing the body

小貼士 / TIPS

桃膠含豐富的植物膠原，有養顏潤燥的功效，是現時流行的養生食材，適合秋冬乾燥季節食用。挑選較少雜質的桃膠為佳。

Peach resin is rich in plant collagen that can help improve complexion and moisten dryness. It is considered a popular health food and is suitable to be consumed during the dry autumn and winter. Peach resin with less impurity is recommended.

材料

桃膠 .... 1 兩
蓮子、乾百合 .... 各 2 兩
紅棗 .... 4 粒
冰糖 .... 2 兩

做法

① 桃膠用水浸一晚，挑去污
垢及雜質，洗淨，飛水，
過冷河，瀝乾水分，備用。

② 蓮子用水浸 2 小時，去芯，
洗淨。

③ 紅棗、百合、冰糖洗淨；
紅棗去核。

④ 湯煲內加入水 10 碗，放入
蓮子煲滾半小時，下百合、
紅棗再煲半小時，加入桃
膠、冰糖煲 10 分鐘即成。

Ingredients

38 g peach resin
75 g each of lotus seeds and
   dried lily bulb
4 red dates
75 g rock sugar

Method

1. Soak peach resin in water
   overnight. Remove the dirt and
   sediments. Rinse peach resin,
   briefly scald with boiling water,
   rinse with cold water, drain.

2. Soak lotus seeds in water for 2
   hours. Remove pith and rinse
   lotus seeds.

3. Rinse red dates, lily bulb and
   rock sugar. Remove seed of
   red dates.

4. Add 10 bowls of water to soup
   pot. Add lotus seeds and boil
   for 1/2 hour. Add lily bulb and
   red dates. Simmer for 1/2 hour.
   Add peach resin and rock
   sugar. Simmer for 10 minutes.
   Serve.

Starfruit and Pear Juice

楊桃雪梨汁

● 清心潤肺、滋潤皮膚 ●

regulating the Heart meridian;
nourishing the Lungs; moisturizing the skin

材料 (份量：300 毫升)

楊桃 .... 2 個
雪梨 .... 2 個

做法

❶ 楊桃洗淨，去核，切碎。
❷ 雪梨洗淨，去芯，切碎。
❸ 將材料放進榨汁機內，榨出汁液，即可享用。

Ingredients (yields 300ml)

2 starfruits
2 Ya-li pears

Method

1. Rinse the starfruits and remove the seeds. Cut them into pieces.
2. Rinse, core and dice the pears.
3. Press all ingredients through an electric juicer. Serve the juice.

小貼士 / TIPS

楊桃及雪梨必須切碎，才放入
榨汁機攪成汁。

You need to cut the starfruits and
Ya-li pears into pieces, then put
them in an electric juicer until
done.

Double-steamed Pear with Chuan Bei and Lily Bulbs

川貝鮮百合燉秋梨

● 潤燥、化痰、止熱咳，紓緩咳嗽 ●

nourishing dryness; expelling phlegm;
stopping Heat-type coughing; easing cough

材料

川貝 ....半兩
鮮百合.....1 包（約 2 至 3 個）
鴨咀梨..........................4 個
冰糖 ....半湯匙

做法

❶ 川貝洗淨，瀝乾水分備用。

❷ 鮮百合切去頭尾兩端，撕成瓣狀，洗淨，用水浸過面。

❸ 鴨咀梨洗淨，切角，去芯。

❹ 將鴨咀梨、冰糖及川貝放於燉盅內，傾入滾水 3.5 杯，加蓋，隔水大火燉 15 分鐘，轉小火燉 45 分鐘，加入鮮百合再燉 15 分鐘即成。

Ingredients

19 g Chuan Bei
1 bag fresh lily bulbs
   (about 2 to 3 pieces)
4 Ya-li pears
1/2 tbsp rock sugar

Method

1. Rinse Chuan Bei. Drain and set aside.
2. Cut off both ends of the lily bulbs. Tear into petals. Rinse and soak in water (enough to cover the surface). Set aside.
3. Rinse the pears. Cut into wedges. Remove the cores.
4. Put the pears, rock sugar and Chuan Bei into a stewing pot. Pour in 3.5 cups of boiling water. Cover with the lid. Double-steam over high heat for 15 minutes. Turn to low heat and double-steam for 45 minutes. Add the lily bulbs and double-steam again for 15 minutes. Serve.

小貼士 / TIPS

寒咳者只要刪掉川貝，用雪梨、
乾百合及陳皮 2/3 個燉吃也可。
For the people who have Cold-type
coughing, it has the same effect on
them by using pear, dried lily bulbs
and 2/3 dried tangerine peel. Do not
use Chuan Bei in this case.

養好肺・強體質

Creamy Walnut Paste with White Fungus

雪耳合桃糊

● 補腦、養顏潤燥 ●

材料

合桃肉............................ 4 兩
雪耳 .... 1 朵
淡奶 ....2 湯匙
冰糖 .... 2 湯匙（舂碎）

粟粉獻

粟粉 ....1 湯匙
水 ....4 湯匙
* 調勻

做法

❶ 雪耳用水浸 2 小時，剪去硬蒂，撕成細朵，飛水 3 分鐘，盛起，過冷河，瀝乾水分。

❷ 合桃洗淨，與水 1.5 杯同放於攪拌機內，磨成合桃漿，用隔篩過濾，即成幼滑的合桃漿。

❸ 煮滾清水 2 杯，放入雪耳及冰糖煮 5 分鐘，傾入合桃漿拌勻煮滾，下粟粉獻邊煮邊拌，轉小火煮至微滾，最後加入淡奶拌勻即成。

小貼士

拌入淡奶令合桃糊帶一陣奶
香味，吃起來口感細緻嫩滑。

CREAMY WALNUT PASTE WITH WHITE FUNGUS

strengthening the brain; improving skin texture; nourishing dryness

Ingredients

150 g shelled walnuts
1 head dried white fungus
2 tbsp evaporated milk
2 tbsp rock sugar (crushed)

Thickening glaze

1 tbsp cornflour
4 tbsp water
* mixed well

Method

1. Soak the white fungus in water for 2 hours. Cut off the hard stalks. Tear into small pieces. Blanch for 3 minutes. Remove. Rinse in cold water. Drain.

2. Rinse the walnuts. Grind with 1.5 cups of water in a blender into walnut batter. Sieve out the creamy walnut batter.

3. Bring 2 cups of water to the boil. Put in the white fungus and rock sugar. Cook for 5 minutes. Pour in the walnut batter. Mix well. Bring to the boil. Add the thickening glaze. Keep stirring while cooking. Turn to low heat and cook until it slightly boils. Finally fold in the evaporated milk to finish. Serve.

TIPS

It is to make the walnut paste sweet-scented, silky and delicate when folding in evaporated milk.

清潤茶飲

Well-being tea

健肺滋潤的日常茶飲，
在平日繁忙的工作，讓自己益補自療！

Sha Shen, Yu Zhu and Mai Dong Tea

沙參玉竹麥冬茶

● 化痰止咳、潤肺 ●

expelling phlegm; stopping cough; nourishing the Lungs

材料

沙參 ....半兩
玉竹 ....半兩
麥冬 .... 1 兩

做法

❶ 沙參、玉竹及麥冬洗淨，
備用。

❷ 全部材料放入瓦煲內，注
入清水 7 杯煲滾，轉小火
煲 1 小時即成。

Ingredients

19 g Sha Shen
19 g Yu Zhu
38 g Mai Dong

Method

1. Rinse Sha Shen, Yu Zhu and
Mai Dong. Set aside.

2. Put all ingredients in a clay pot.
Pour in 7 cups of water and
bring to boil. Turn to low heat
and simmer for 1 hour. Serve.

小貼士 / TIPS

沙參化痰止咳；玉竹潤燥通便；麥冬潤肺除煩。此茶滋養肝胃、利咽生津。

Sha Shen expels phlegm and stops coughing. Yu Zhu nourishes the body and eases constipation. Mai Dong nourishes the Lungs and calms the nerves. This tea helps to nourish the throat and stimulates body fluid secretion, it is good for the Liver and Stomach.

養好肺・強體質

Lotus Root Knot and Pear Tea with Couch Grass Root

藕節雪梨茅根茶

● **清熱解燥、潤肺** ●

clearing Heat; moistening dryness; nourishing the Lungs

小貼士 / TIPS

藕節是蓮藕節端部分曬乾而成，中藥店有售。

Lotus root knot has been sundried and it is sold in Chinese herbal shops.

材料

藕節 .... 2 兩
雪梨 .... 4 個
鮮茅根 .... 1 紮
冰糖 .... 2 粒

做法

❶ 雪梨洗淨外皮，切成 4 件，
去蒂、去籽。

❷ 藕節、茅根一同洗淨。

❸ 所有材料（冰糖除外）放
入煲內，加入清水 8 杯煲
滾，轉小火煲半小時，下
冰糖煮溶，待暖飲用。

Ingredients

75 g lotus root knots

4 Ya-li pear

1 bundle fresh couch grass root

2 cubes rock sugar

Method

1. Rinse the pear and cut into
   quarters. Discard stem and
   seeds.
2. Rinse lotus root knots and
   couch grass root.
3. Add all ingredients to pot
   (except rock sugar). Add 8
   cups of water. Bring to boil.
   Adjust to low heat and simmer
   for 1/2 hour. Add rock sugar
   until dissolves. Serve warm.

養好肺 • 強體質

Ya-li Pear Tea with Fresh Lily Bulb

鮮百合雪梨茶

● 清熱生津、潤肺止咳 ●

expelling Heat; stimulating body fluid secretion;
nourishing the Lungs; stopping cough

材料

鮮百合 .... 2 球
雪梨 .... 4 個
冰糖 .... 1 粒

Ingredients

2 heads fresh lily bulbs
4 Ya-li pears
1 cube rock sugar

做法

❶ 鮮百合切去頭尾兩端的焦
黑部分，撕成瓣狀，洗淨。

❷ 雪梨洗淨外皮，切角，去
芯。

❸ 煲內注入清水 8 杯煲滾，
加入雪梨，用中小火煲 40
分鐘，下冰糖及鮮百合再
煲 10 分鐘即可。

Method

1. Cut off the head, root and the
   black parts from lily bulbs. Tear
   into small pieces. Rinse.
2. Rinse pears. Cut into wedges
   and remove the cores.
3. Bring 8 cups of water to boil.
   Put in pears. Boil over medium-
   low heat for 40 minutes. Put in
   rock sugar and lily bulbs and
   boil for 10 minutes. Serve.

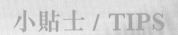

小貼士 / TIPS

經煲煮的雪梨不太寒涼，加上百合同煲，有清熱及滋潤肺部之效。

Cooked pears are not too cold in nature. When cooked with lily bulbs, they can expel Heat and nourish the Lungs.

養好肺・強體質

Dried Tangerine Peel and Gold Luo Han Guo Tea

陳皮金羅漢果茶

● 清肺利咽、順氣化痰 ●

clearing the Lungs; soothing throat;
improving breathing; dissolving phlegm

材料

金羅漢果 .... 1 個
陳皮 .... 2/3 個

Ingredients

1 gold Luo Han Guo
2/3 dried tangerine peel

做法

❶ 陳皮用水浸 1 小時，刮淨
內瓤，洗淨。

❷ 金羅漢果洗淨，切成 4 塊。

❸ 湯煲內加入清水 7 碗，放
入金羅漢果、陳皮煲滾，
轉小火煲 15 分鐘，熄火，
加蓋待半小時即可飲用。

Method

1. Soak dried tangerine peel in
   water for 1 hour. Scrape off the
   peel and rinse.
2. Rinse gold Luo Han Guo and
   cut into quarter.
3. Add 7 bowls of water in soup
   pot. Add gold Luo Han Guo
   and dried tangerine peel. Bring
   to boil. Adjust to low heat and
   simmer for 15 minutes. Turn off
   the heat. Cover the lid and let
   stand for 1/2 hour. Serve.

小貼士 / TIPS

金羅漢果營養價值很高，清熱、潤肺、止咳，對喉嚨咽喉有一定幫助，適合肺熱及肺燥咳嗽人士使用。

Gold Luo Han Guo has high nutrient value, helps clear Heat, moisten the Lungs, relieve cough and soothe throat, beneficial to those suffering from Dryness-Heat in the Lungs induced cough.

養好肺・強體質
64 道 增 強 肺 臟 日 常 食 療

主編
萬里編輯委員會

責任編輯
簡詠怡

美術設計
鍾啟善

排版
劉葉青

出版者
萬里機構出版有限公司
香港北角英皇道499號北角工業大廈20樓
電話：2564 7511　傳真：2565 5539
電郵：info@wanlibk.com
網址：http://www.wanlibk.com
　　　http://www.facebook.com/wanlibk

發行者
香港聯合書刊物流有限公司
香港新界大埔汀麗路 36 號
中華商務印刷大廈 3 字樓
電話：2150 2100　傳真：2407 3062
電郵：info@suplogistics.com.hk

承印者
中華商務彩色印刷有限公司
香港新界大埔汀麗路 36 號

出版日期
二零二零年四月第一次印刷

ISBN 978-962-14-7218-2